Thomas① Hungerford (founders monument in Hartford, Ct.)
b. in Eng. d. 1663 Hadlyme Ct. went to New London, Ct.
with 11 others ~~to start~~ to help Winthrop in
1652 cleared land where Ft. Trumbull now stands (*) in
m. #1 (maybe a Sarah Green) d. 1658 had Thomas② + Sarah
#2 Hannah Willey b 1642 — had Hannah

Thomas② Hungerford (a blacksmith) b. 1648 Hartford / New London Hadlyme
d. Jan 29, 1713 age 65 — Cove Burying Ground — above E. Haddam landing
m. — Narraganset War (Eliz. b. 1670) to Mary (b.1659 ? Gray?)

John③ Hungerford b 1672 near New London Ct. d. July 9, 1748 Hadlyme Chur.
inherited homestead near corner of Bone Mill Rd. + Hemlock Valley Rd. Cemetery
m. Dec. 3, 1702 (age 28)(East Haddam Records Vol I p4) to Deborah Spencer
(of Timothy, Ensign Gerard) 1683 - 1750 buried Hadlyme Cong. Church Cemetery

Robert④ Hungerford 1716 - 1794 (Hadlyme) on the hill ½ mi. N.W. of Cong. Church
m. 1736 Grace Holmes b 1717 to Cpt. John Holmes + Mary Willey (John + America moore)

Robert⑤ Hungerford b 1755 - 1834 m #1 Louise Warner d. 1777 (soon
after birth of son Robert while husband at Lexington + Concord) buried
in Seldon Cemetery (#2 her cousin Olive

Robert⑥ Hungerford b. Jan. 17, 1777 Hadlyme d. 1842
m. Dec. 28, 1820 (after building new house that later was
purchased for Cong. parsonage + after fire rebuilt on same
to Huldah Riley Skinner 1796 - 1868 (#2 Samuel Brooks)

E.C.① Edward⑦ Codrington Hungerford 1837 - 1910 buried Laurel C. Chester Ct.
m. 1859 Chester Irma Gilbert Daniels 1834 - 1930 of Col Charles + niece
(b. in the Greek Revival House on Liberty St. in Chester, Ct. 3 Gilbert

Dr. Robert⑧ Hungerford 1862 - 1888 d. while visiting wifes family in Canada.
m. 1885 to Alice Maria Abbott 1860 - 1935 (#2 Frank Nalder) buried
in Walla Walla, Wa. Cemetery but died in Portland, Oregon dau. of
Civil War Chaplain Thomas Jefferson Abbott + Eliz. (Arnold) Moulton

Robert⑨ Abbott Hungerford 1887 Chester Ct. - 1959 Portland, Ore.
m. 1911 Lewiston Ida. Mary Alice Small dau. of Ira Ambrose Small
and Alice Clara Chamberlain (of Rev. Pearl S. Chamberlain + Alice Eaton Ab.

Robert⑩ Ira Hungerford b 1921 Portland Or. m 1948 Elizabeth Hansen
b. 1935 Aurora, Nebraska to C.C. Hansen + Edith V. Nilson (of Eric, Huldah N.

FLAMING BRANDS

Fifty Years of Iron Making
in the
Upper Peninsula of Michigan
1848 - 1898

By
Kenneth D. LaFayette

this book p 27, p 91
p 28,

Edited by
Thomas A. Johnston,
Burton H. Boyum, and
James L. Carter

Kenneth D LaFayette

i

Library of Congress Card Catalogue Number 77-072800

International Standard Book Number 0-918616-01-8

Printed in the USA

To Janice, Allan, Todd, Gregory, and Jason.

For 'tis iron that dwells in the blood of the heart,
That adds to her beauty, and enlivens the bliss
Of the lover's caress, and the pure friendly kiss.

— from "Iron the Precious Metal"
as quoted in *The Iron Agitator,*
Ishpeming, Michigan
November 27, 1886

ABOUT THE AUTHOR

Kenneth LaFayette is a native of the rural Marquette County village of Sands. In 1937 he moved with his family to Marquette where he spent most of his youth and graduated from high school. He served two years with the U.S. Army in France, and returned to Marquette where he married Janice Hawes. They are the parents of four sons.

LaFayette has attended Northern Michigan University and is a dispatcher for the Upper Peninsula Power Company at its System Control Center in Ishpeming.

The author has deep roots in the Upper Peninsula. Originally from the Montreal area of Quebec, his family came to Marquette County from Vermont in 1911, settling in Harvey where his grandfather, Edward LaFayette, built a mill for the manufacture of wooden barrel hoops from black ash logs. The author's maternal ancestors, the Johnsons, came to Ironwood from Sweden in 1899, moving to Negaunee in 1903 where Charles Johnson, his grandfather, worked in the Athens and Negaunee Mines. The family later established a farm in Skandia Township of Marquette County.

LaFayette has developed an avid interest in Upper Peninsula history. What began as a hobby has become a serious avocation. *Flaming Brands* is his first publication.

He is a member of the Marquette County Historical Society and the Marquette Range Engineers.

PREFACE

It is a distinct pleasure for The Cleveland-Cliffs Iron Company to sponsor the publication of the first edition of *Flaming Brands*. We applaud the author for his diligent efforts and accurate portrayal of this fascinating subject.

In his colorful story, Mr. Kenneth LaFayette traces the history of charcoal iron production from its earliest origins in the Upper Peninsula, providing added perspective into Michigan's iron mining industry.

Cleveland-Cliffs is closely linked to charcoal iron production through The Jackson Iron Company which first produced charcoal iron at its Carp River Forge on February 10, 1848, and later at its Fayette Iron Works. In addition, the predecessor companies of Cleveland-Cliffs — The Cleveland Iron Mining Company and The Iron Cliffs Company — both ventured into charcoal iron production in their early days.

Mr. LaFayette's treatise brings to mind another scholarly study of charcoal iron production which was authored by William G. Mather, president of Cleveland-Cliffs from 1890 to 1933, and published in the 1903 *Transactions of the Lake Superior Mining Institute*. Mr. Mather recognized, as we do today, the great contribution this facet of the iron industry has made not only to the Upper Peninsula of Michigan, but to the nation as a whole.

This remarkable work, resulting from much reading and research by Mr. LaFayette, reminds us again of our rich heritage and the important role these early pioneers played in laying the groundwork for today's iron and steel industry.

SAMUEL K. SCOVIL
President and Chief Executive Officer
The Cleveland-Cliffs Iron Company

Cleveland, Ohio
January 12, 1977

CONTENTS

ILLUSTRATIONS

ACKNOWLEDGEMENTS

Many people have given generously of their time in assisting me during several years of research on the charcoal iron industry of the Upper Peninsula, and in compiling this volume.

I am deeply appreciative of the assistance given by Esther B. Bystrom, executive director of the Marquette County Historical Society, who has encouraged me from the time I first began working in the Society's excellent John M. Longyear Research Library. Thanks also go to Joan P. Weesen, the Society's secretary-treasurer, and to Shirley Peano, Lydia Werner and Edna Paulson, members of the Longyear Library staff, for their aid in locating material and their helpful comments. Wesley Perron, past president of the Society, was very generous of his time in providing information on Upper Peninsula railroads.

The Peter White Public Library, with its array of historical works and micro-filmed papers, was an invaluable source of material. My thanks go to Ruth Kell, head librarian, and to the members of her staff for their help — and patience. Dorothy Constance was especially helpful in finding local blast furnace information.

Newspaper files at the Carnegie Public Library in Ishpeming and the Negaunee Public Library also provided much useful data.

Comments of Dr. Richard O'Dell, retired professor of history at Northern Michigan University, and Dr. Russell Magnaghi, member of Northern's history department faculty, were very helpful in organizing the material for publication. The suggestions of Dr. Richard P. Sonderegger, a veteran member of the NMU faculty, aided materially in developing the Introduction.

I especially want to thank Mr. James L. Carter, director of the Northern Michigan University Press, for his encouragement from the first draft to the final edited copy; his faith in the project has been an inspiration.

I am grateful to Sergeant Thomas A. Johnston of K.I. Sawyer Air Force Base, a former graduate assistant in history at NMU, who spent many hours editing the manuscript, and to Mr. Burton Boyum, director of administrative affairs for The Cleveland-Cliffs Iron Company, for his editorial assitance from the viewpoint of the mining industry.

And lastly, to my wife Janice, goes my deepest appreciation for her patience and understanding during the many hours spent in researching and writing this book.

Kenneth D. LaFayette

Marquette, Michigan
January 20, 1977

INTRODUCTION

Throughout the United States in the nineteenth century, changing technology in iron making steadily brought about innovations in its manufacture. This was primarily due to changes in its use, improvements in transportation, and the demands of an ever-growing industry to provide for the expanding population — but not necessarily in that order. To understand the advancement made in the iron industry, one must realize that it was not until in the 1830s that the first coal-iron smelting furnace in the United States was put into limited operation.

The attempts at making iron in forges in Marquette County from 1848 until 1862 were brave undertakings, encouraged by the quantity and quality of iron ore and hardwoods in the area. But the lack of facilities for shipping products to the lower lakes markets hindered the industry for many years. It was not until the Sault Canal was completed in 1855 that any thought could be given to developing an iron industry of any size near the ore reserves.

During the Civil War many of the Southern States' furnaces had become outdated and were abandoned.[1] The Civil War had a great influence on the iron industry when demands for iron increased substantially and continued for some years after. This revived hope for the investors in Upper Peninsula iron works[2] also, new blast furnaces were built in many advantageous places in Marquette County. Up to this time all the furnaces in the Peninsula were run on charcoal, except the anthracite coal burning Northern Furnace at Harvey. But this, too, was soon won over by the surrounding hardwood forests.

In England the charcoal fired furnaces had used up practically all the hardwood forests by the eighteenth century. Large reserves of coal were available, and from the coking ovens came fuel which made a new iron having different properties than charcoal iron. Although coke iron soon became widely used, charcoal iron was still produced for those who preferred it.[3]

The big iron producing centers in the United States had the same problems early in the nineteenth century, and iron makers had to look for a more dependable supply of fuel. With forest lands rapidly falling under the axe near the iron works, transportation costs increased for hauling charcoal from the supplier to the user. As in England, coal gradually replaced charcoal as a fuel.

Pennsylvania had many natural advantages that helped make Pittsburgh the "iron city" of America. The Great Lakes waterways provided a cheap route to the industrial cities on her shores, huge beds of bituminous

---

[1]Schallenberg, Richard H. *Technological Innovation in the American Charcoal Industry 1830-1930*. 1970, p. 34.
[2]*Geological Survey of Michigan, 1873*, p. 34. New York, Julius Bien, 1873.
[3]Schallenberg, p.10.

coal for coking were within her boundaries, and extensive ore reserves allowed her to produce iron in large amounts — cheaper than in many places. Coke iron was first made in the United States at the Mary Ann Furnace in Huntingdon County, Pennsylvania, in 1835, and the following year anthracite coal was used for the first time to make iron at the Lucy Furnace near Pottsville.

Charcoal iron led production until 1854. The following year, the percentage of charcoal iron in the total output of iron in the United States began to decline. Anthracite iron production led charcoal iron for the first time in 1855, and in 1875 coke iron took the lead.

One of the requirements for making charcoal iron was the availability of immense quantities of hardwood near the iron works. With vast forests of hardwood at hand in the Upper Peninsula, it was only natural for the iron companies to make forges and blast furnaces that used charcoal. (The large amount of fuel that these iron works used could only be delivered to the south shore of Lake Superior at great expense.)

The last forge built in the Peninsula came into use in 1855, the same year the Sault Canal was built. After that, the many new iron companies starting here constructed the much larger blast furnaces which required hundreds of tons of castings and machinery. Although blast furnaces had been used in America for many years, it was too costly before the canal was built to ship and portage all of the machinery required by such furnaces. As it was, the smaller forges were ideal for the existing conditions. At Marquette there were no roads to speak of, and the only time ore could be moved from the mines to the forges was in the winter when packed snow roads made easy sledding. Charcoal iron was still very popular when these forges were built, as the coal and coke iron industries were then in their infancy.

The forges erected in the Peninsula to smelt the iron ore were little changed from the Catalan forges developed in northern Spain about the eighth century, and the wrought iron made in them was still a very popular product.[4] Because of its low carbon content and fusability, wrought iron was preferred by the blacksmiths and others who wanted a strong, malleable iron which could absorb shock, for making such things as nails, tools, horseshoes, and wagon wheel rims. Cast iron from the blast furnace was brittle from the high carbon content and unless it had further treatment to remove impurities, it was unfit for these uses.

The construction of the forges was more open than for blast furnaces and, although large amounts of charcoal were used in them to smelt the ore (much more per ton than in a blast furnace), temperatures could not be raised high enough to allow the iron to melt and absorb carbon from the fuel. Therefore, the iron made in them was almost carbon-free. Later charcoal blast furnaces produced cast iron with a carbon content of

[4]Fisher, Douglas Alan, *The Epic of Steel* (New York, 1963), p. 26.

approximately 3½ percent. The higher stack temperatures due to the larger hot blast and enclosed fire, caused the iron to melt and take on the carbon released from the charcoal.

At the Northern, Grace, and Marquette & Pacific Furnaces where coal and/or coke were used for fuel, sulphur and phosphorus, with carbon and other brittling elements, were absorbed in the melted iron. Resmelting of the pigs in puddling furnaces, air furnaces, or cupolas removed most of these elements and rendered the iron malleable and heat treatable. After the impurities were burned off in this resmelting, the iron could then be used in malleable iron castings such as railroad car wheels, rails and many items where durability was desired. Unrefined cast iron from the blast furnaces was used in certain items such as stoves and kettles where brittleness was not detrimental to the finished product and costs were important. At the Menominee, Gogebic and Morgan iron works, molten iron was tapped from the furnace and run into moulds making cast iron items for their own use — a practice probably carried out at all iron works.

The making of wrought iron in the Upper Peninsula from cast pig iron was tried at different times with varying amounts of success. C. Donkersley's experiments with a rotary puddling furnace of his own design at the Morgan Furnace ended in failure, but some "steel" was made and converted into fine grained wrought iron hammered bars. The furnace was supposed to puddle a ball of iron weighing 1,000 pounds. The Marquette & Pacific Rolling Mill furnace had a group of seven puddling and heating furnaces that would not work with bituminous coal, but when switched to coke they operated well, and many tons of cast iron were refined into muck bar and merchant bar iron. At the Fayette Furnace in Delta County, a cupola was erected and with it was produced iron for the furnace's own "chills" and other castings. Others probably existed in the Peninsula but no record could be found of them.

The fluxing materials used in the local forges and furnaces were in great supply, and the limestone and dolomite that abounds throughout Marquette County was used with good results. Limestone was the ideal flux for iron making, and at places like Fayette where native limestone was at hand, it was used in the furnace. Some furnace companies in Marquette County had limestone brought in by boat — the Marquette & Pacific for one — and others imported Kelly Island limestone from Lake Erie, while the Menominee Furnace at times used slacked lime.

A flux is necessary in the iron making process to melt and gather all of the silica present in the iron ore. When melted, the slag formed is lighter than the molten iron and floats on top. A special opening or "notch" was present on the furnace and through it the slag was drawn off for disposal.

The only experiment for using peat as a fuel in the Upper Peninsula was at Ishpeming, where a large peat bed is found. The peat was ground up and dried before use and, being light and bulky, large amounts were required for each charge because of its rapid consumption and low calorific value. Varying amounts of charcoal were mixed with the peat and at times success seemed near. But when peat was used by itself as the fuel, the outcome was a dismal failure.

Prior to 1844 when iron ore was found at Negaunee, nearly all iron produced in the United States was made with charcoal. A few years later, in 1860, only 30 percent of the total was charcoal iron; by this time the making of iron with hard and soft coal and coke was greatly increasing. By 1880 the total production of all iron in the United States had increased to over 4,295,000 tons; of this 1,807,000 tons were made with anthracite coal, 1,950,000 tons with bituminous coal and coke, and 537,558 tons using charcoal.[5]

As the demand for iron increased, the uses for charcoal iron decreased. From a country dependent on small foundries and the local blacksmith for iron goods, the introduction of coke iron making allowed greater expansion of the industry — due to the larger output of a less expensive product. No longer did the nation have to pay for the increasing cost of charcoal iron because of dwindling forests. The huge reserves of bituminous coal assured a continuous supply of the cheaper fuel — coke.

The low sulphur content of charcoal iron made it a favorite with U.S. foundries for making railroad car wheels, and into the twentieth century it was used by many of them exclusively. But in later years the steel wheel was used because of heavier loads and higher speeds. The Martel Furnace at St. Ignace was erected by a railroad car wheel concern of Erie, Pennsylvania, to provide the special charcoal iron for wheels.

As early as 1858, there were two foundries in Marquette using the locally made pig iron for castings. At the foundries the pigs were resmelted in furnaces to burn off the impurities for proper heat treating and wear. Orders were filled for the Lake Superior copper mining companies. At this time, wagon "skeins" (or castings) were in demand and were also being cast regularly. The Lake Superior Foundry cast sections three feet by four feet which were assembled into the first cast iron charcoal kiln in the Upper Peninsula in February of 1859. Two of these, measuring 24 feet in diameter and cone shaped, were erected at the Pioneer Furnace that year and more were ordered. By March of 1859, C. Donkersley's foundry had produced more than 150 railroad car wheels for the Bay de Noquet & Marquette Railroad.

The Bessemer process of making steel was patented in the United States by Sir Henry Bessemer of England in 1856. This method involved blowing air through or over the molten iron in converters to burn off the harmful metaloids. But, in the first 20 years of steel making in this manner,

---

[5]*Mining Journal*, Marquette, Mich., (Weekly Edition), March 12, 1881.

low phosphorus pig iron had to be used as the converters would not eliminate this element. In 1877, two Englishmen, Sidney Thomas and P. C. Gilchrist, found that by lining the converters with limestone or dolomite instead of sandstone or clay, the phosphorus would pass into the slag or burn off. However, this method was never used in the United States as the basic open hearth furnace later replaced it.

Besides a variety of iron ores, different grades of foundry iron and two grades of charcoal pig iron (made especially for the Bessemer converters in the local blast furnaces), Marquette County also supplied a lining material for the converters as noted in the May 19, 1877, issue of *The Mining Journal:*

> H. A. Burt, of the Peninsula Iron Co., has sold 3,000 tons of the quartz from his mine on Carp River, and has let a contract to Hursley & Powell to get it out. They are now at work, and have mined already some 700 or 800 tons. The *City of Duluth* took down 150 tons on her last trip destined for the Vulcan Iron Works at St. Louis, Mo. This material is used for lining the converters in Bessemer steel works, for 'fix' in puddling furnaces, & c. . . This substance was formerly imported almost exclusively from England . . . .

The high cost of shipping coke to the Upper Peninsula and getting the pig iron to the industrial markets made iron making uneconomical in the nineteenth century. The shipping of ore to the coke and coal furnaces was expanded when the Sault Canal was completed, and nearly every year thereafter the tonnage increased. Plans for making Marquette a great iron-making city were talked of at different times but were never pursued. The arguments against shipping huge amounts of coke to the Lake Superior region always won out. While coke was light and easily handled, it was bulky and tended to deteriorate after a time. Being crushed and broken during shipment slowed its burning abilities and the absorption of moisture ruined some of its qualities. On the other hand, ore could be handled, dumped, stockpiled, shoveled and dumped again unharmed.

Upper Peninsula furnaces tried them all — charcoal, anthracite and bituminous coal, peat, coke, charcoal mixed with pine, and at the Morgan Furnace a brief encounter with oil — but charcoal remained the most used fuel.

An early view of Marquette, from Foster & Whitney's *Report on the Geology of the Lake Superior Land District, II,* 1851.

Chapter I

THE FIRST IRON MADE
FROM LAKE SUPERIOR ORE

The Upper Peninsula was covered with a dense, virgin forest in 1844 with vast, untouched mineral deposits lying beneath it. United States Deputy Surveyor William A. Burt and his party discovered iron ore near Teal Lake in Marquette County on September 19 of that year, when the ore caused variations in the direction of the needle of his solar compass. Knowledge of Burt's discovery was not widely known and it was not until the following year that this important discovery was exploited.

An earlier survey of the Upper Peninsula, conducted by State Geologist Dr. Douglass Houghton, had revealed the presence of copper and silver deposits in the Keweenaw Peninsula. Philo M. Everett of Jackson, Michigan, heard of the discoveries and, with S. T. Carr, W. H. Monroe, and E. S. Rockwell, formed the Jackson Mining Company to tap these valuable minerals. The party left Jackson on July 23, 1845, to prospect in the Keweenaw region.

While passing through Sault Sainte Marie, Everett learned of Burt's iron ore discovery at Teal Lake. When his party failed to find silver or copper on the Keweenaw Peninsula they decided to search for Burt's iron ore discovery. The party stopped at a small Indian village near the mouth of the Carp River and persuaded a local Indian, Marji-gesik, to aid them in their search. He escorted them to the general area of the glistening "heavy rock" and the deposit was found. Shortly after, the area was named Jackson Mountain.

The company claimed an unsurveyed area one mile square for its location on a permit issued by the Secretary of War to a James Ganson, and later when the tract was surveyed, it purchased the land for $2.50 an acre.[1]

In 1846, Abram V. Berry, the first president of the Jackson Mining Company, and others secured 300 pounds of the ore. They eventually had it made into a bar of iron by Aaron K. Olds, at the Branch County Iron Works in southern Michigan.[2] During the winter of 1846-47, the dozen or so persons remaining at the Jackson Location cut a road from the shore of Lake Superior to a site they had selected on Carp River, lying three miles east of Jackson Mountain. Here, they erected a few log houses, a dam and cut some timber for the forge building.

Berry, in the meantime, was diligently at work in Jackson acquiring the bellows and machinery necessary to build the forge on Carp River, to make charcoal bloom iron from the rich iron ore. On July 7, 1847,

[1]*Mining Journal*, July 6, 1872.
[2]Michigan History Division, *Carp River Forge: A Report*, 1975, p. 5.

the mechanical parts of the first iron smelter on the Upper Peninsula arrived on board the steamer *Independence*,[3] accompanied by Ariel N. Barney, his son, Samuel, Aaron Olds, William Lemm, Carr, and others, to make a total of 24 men and two women. They all stayed in one large log house at the Carp River site during the winter of 1848. The *Independence* delivered the first yoke of oxen to be used in Marquette County and it was Samuel Barney's job to drive them.

A water-powered sawmill was put up at the site and, on the tenth of February, 1848, Ariel Barney made the first bloom of iron in the Carp River Forge.[4] Unfortunately, the dam proved inadequate and in March an early "freshet" washed it out, doing considerable damage to their water power. The dam provided water through a wooden flume that turned the large water wheel located in a low lying area approximately 25 feet east of the forge house.[5] Early that summer Everett returned to the Carp River settlement and supervised repairs to the 18-foot-high dam. With water power restored the forge was again put into operation turning out high grade blooms of iron.

The first record found of bloom (or bar) iron shipped from the Carp River Forge — and from the iron district of Upper Michigan — is recorded in the *Lake Superior News* of Sault Sainte Marie in its issue of May 13, 1848:

> "A small boat, which coasted down from Carp River, week before last, brought down from the Jackson Iron Works at that place, some six or seven hundred weight of bar iron, manufactured there . . . of very superior quality . . . ."

Ariel and Samuel Barney and three others were in the boat bringing the shipment which included the first iron made in the forge — to the Sault to be sent on to Jackson for display.

With the loss of the dam in March an additional hardship was put on the small group for it had also provided the motive power for the saw mill. As noted in the weekly *Mining Journal* of July 14, 1900, a shortage of food developed shortly after and the only way to relieve the situation was to reach Abraham Williams' trading post on Grand Island, a distance of about 40 miles on Lake Superior, but they lacked a boat. The men proceeded to whipsaw lumber and built a boat large enough to bring supplies from Grand Island and then used it to transport the five men and iron to Sault Sainte Marie.

The forge at first was built to use a cold air blast to fan the flame. This was later changed to a hot blast, which saved both time and fuel.[6] The air for the hot blast was heated by being forced through heavy cast iron pipes running through the center of the stack where they were heated by

[3]*Lake Superior News and Mining Journal*, Sault Ste. Marie, July 24, 1847.
[4]*Lake Superior Journal*, Sault Ste. Marie, June 25, 1851.
[5]The resting place for the water wheel was found in 1974 by archaeologists working for the State of Michigan History Division in their attempt to locate the various parts of the iron works for the Carp River Forge Association.
[6]*Lake Superior Journal*, July 9, 1851.

the hot gases from the charcoal fuel. The air was furnished through the use of bellows which were actuated by a wooden shaft running directly from the water wheel. One would find on entering the forge house that "There were upon either side of the stone arch, and arranged opposite each other, four fires . . . ."[7] The forge at different times in its short life used one to eight fires, from which the blooms of iron would be taken about every six hours and shaped into bars under the heavy trip hammer.[8] By May of 1848, many thousands of dollars were invested in the enterprise. On June 6, 1848, the Jackson Mining Company was organized with a capital stock of $300,000 in shares of $100 each, and on April 2, 1849, the charter was amended changing the name to the Jackson Iron Company.

In April of 1850, the Jackson Company's forge had two fires in operation turning out 2,000 pounds of bloom iron a day at an expense of just over $20.00 per ton.[9] Later that spring, however, the company was ready to shut the fires down as the iron business was not paying its way and their debts were growing rapidly. To make matters worse, the unpaid, unhappy men were circulating a rumor that Ezra "Czar" Jones, then president of the company, could well be the victim of a hanging. Fearing for his life, Jones later hired young Peter White to guide him south through the dense forests to Bay de Noc and safety.

THE MARQUETTE IRON COMPANY

The second forge built in the Upper Peninsula was erected by the Marquette Iron Company in 1849-50, and was located just south of East Baraga Avenue in Marquette, between the outcrop of rock on which the statue of Father Marquette now stands and Lake Superior. Both Amos Rogers Harlow and Robert J. Graveraet have been given credit for organization of the company, which filed articles on May 5, 1849, capitalized at $150,000.

Waterman A. Fisher, the main financial backer of the group with Edward B. Clark and Harlow were all from Worcester, Massachusetts, and they named the site after their home town. Graveraet came from Mackinac Island. These four comprised the Marquette Iron Company.

Graveraet arrived at the site on May 17, 1849, with ten men from Mackinac Island, one of whom was Peter White, to provide labor at the future iron works and on the company's iron ore claim. Harlow, who owned a small machine shop in Massachusetts, constructed and assembled the iron works machinery at a cost of $8,000, and then accompanied it to the Lake Superior country.

When the small schooner *Fur Trader* arrived at what is now Marquette on July 6, 1849, carrying Harlow and his entourage, there was no dock on which to land the forge and sawmill machinery. The schooner was tied

---

[7] *Geological Survey of Michigan*, 1873, p.17.
[8] Swank, James M., *History of the Manufacture of Iron in all Ages* (Philadelphia, 1892), p. 322.
[9] *Lake Superior Journal*, May 1, 1850.

up to what was later called Ripley's Rock about 1,000 feet off shore and there the steam engine, heavy castings and other machinery were unloaded. The equipment was then moved to shore on a slide track constructed for that purpose, but the boilers were plugged at each end and floated ashore from the boat.

A sawmill with a circular saw and powered by a 32-horsepower steam engine was erected and put into operation in October of 1849, to supply lumber for the forge buildings. The forge bellows and trip hammers were also powered by steam, which was supplied through the use of boilers.

Four fires of the Marquette Iron Company's forge were lit on July 6, 1850, and, with ore that had been sledded down from the Jackson Mine the previous winter, they commenced making iron. This forge, like the Carp River Forge, at first used a cold air blast in the fire, but it was soon changed to receive the beneficial hot air blast. The small amount of ore they did have stockpiled was speedily consumed. Another problem which would plague charcoal iron furnaces in the Upper Peninsula into the twentieth century was the constant shortage of charcoal. The forges operated similarly to the ancient Spanish Catalan type and required nearly 200 bushels of charcoal to produce one ton of iron blooms.

In the early years of iron making at Carp River and Marquette, the charcoal was all made in pits as the stone and brick kilns were not built until later. The charcoal pit method consisted of piling four-foot pieces of hardwood to form a mound containing from 25 to 30 cords. This was covered with a layer of small, dry branches called "lapwood," then a layer of wet leaves, and finally a layer of four to six inches of earth dug from a trench around the base of the mound.[10] After covering, the wood was fired and left to char slowly. In about a week or more the finished charcoal was raked away leaving the remainder to simmer until the whole mound was reduced to charcoal. It was then loaded into wagons or sleighs and hauled to the iron works. A pit of this size could yield over 1,000 bushels of charcoal.

By the fall of 1850, the site of the Marquette Iron Company had developed substantially. The company employed 70 men, and utilized oxen and nine five-horse teams to haul ore and iron.

With all of the time, money and labor expended at Carp River to make iron, the Jackson Company did not produce over 100 tons of blooms. Consequently, in the fall of 1850, it leased the forge to two brothers from Columbus, Ohio — Watt and Benjamin Eaton. The Eatons and their men arrived at Carp River on the *Napoleon*, in the middle of December, but had to leave most of their horses and supplies at Sault Sainte Marie until the following spring. After carting their machinery and goods to the iron works, some of the men continued on to the Jackson Mine and that winter broke up and piled over 1,000 tons of ore for themselves and several

---

[10]Trenches of this nature can be found in the woods surrounding Forestville, with charcoal still in the top soil.

hundred tons for the Marquette Iron Company. In March of 1851, the Carp River Forge was making iron again at the rate of about 20 tons per week, having nearly 5,000 bushels of charcoal on hand.

MAKING IRON ON CARP RIVER

On June 25, 1851, the *Lake Superior Journal* in Sault Sainte Marie described iron making at Carp River Forge:

"The ore is first thrown into a large kiln, on a layer of wood, and burned for several days, being rendered by this process brittle. It is then taken to the stamps and pounded quite fine, in which state it is ready for the fire. The furnaces are something like a blacksmith's fire, on a large scale, being open in front and back, enclosed at the sides and tops with heavy cast iron plates, receiving the hotblast at either side . . . .

"The finely pounded ore, together with the charcoal, is thrown upon the fires, in small quantities at a time, in their proper proportions; and after being subjected to this constant heat for three or four hours, the hot air is shut off, and the bulky, shapeless mass of burning iron, streaming with melted earthy substance, is pried out of the fire by the modern Vulcans, and tumbled along to the huge hammer. This mass, weighing about 300 pounds, will not melt with all the heat that can be applied to it under these circumstances; it is of too good a quality for that; only poor article of iron ore is melted in making it into iron; but the foreign matter becomes liquid and collects in and at the bottom of the mass, and is run off into cinders from time to time during the process of heating, and is perfectly driven out under the heavy hammer.

"It is quite surprising to see how easily a mass of this size can be handled by a bloomer, with his heavy tongs. By balancing the tongs, firmly gripping the iron in a loop of suspended chain, he turns it and moves it backwards and forwards with all the precision and ease with which a blacksmith handles a common bar of iron. One of the blooms when finished weighs about 240 pounds, and on account of its unwieldly size and weight, it is sometimes cut in two pieces, being about two feet long and four inches square. In this shape they are sent to market, ready for the rolling mill . . . ."

The Eaton Company produced 200 tons of iron by July of 1851, and almost daily big teams were seen hauling the iron to the lake.[11] Their works utilized about 40 horses with two pairs of oxen, and employed more than 40 men. It also included a store, sawmill and several houses. The combined amount of iron blooms shipped by the two iron companies for 1851 was 500 tons, valued at $50.00 a ton — which amounted to $25,000 for the financially pressed companies.

---

[11]*Lake Superior Journal*, September 3, 1851.

The Sharon Iron Company of Mercer County, Pennsylvania, bought controlling interest in the Jackson Company, including the Carp River Forge, in September of 1851. The Marquette Forge was purchased in May of 1853 by the Cleveland Company, which spent the entire winter building up a stockpile of ore.

C. H. SCHAFFER,

Manufacturer of

Charcoal for Furnace Purposes.

ONOTA, ALGER CO., MICH.

From *Handbook & Guide to Ishpeming, L.., Mich.*, 1886.

Chapter II

THE MARQUETTE FORGE BURNS AND TWO NEW FORGES ARE BEGUN

Ships approaching the Marquette harbor at night could see the bright fires of the Marquette forge at a great distance and, coming into the bay, the loud slamming of the huge forge hammers and roaring of the powerful steam engines echoed over the water. Visitors to the area could go to the iron works, with smoke-belching stacks, and watch the bloomers and workmen tirelessly handle the red hot masses of iron.[1] Since the Cleveland Iron Mining Company bought out the forge it had operated quite steadily, and in the fall of 1853 a large amount of blooms had been made.

On December 14, 1853, a remarkably warm day for that time of the year, Azel Lathrop was busily engaged in the forge, with ". . . a loop under the hammer, shingling, when the alarm of fire was given . . . ." The blaze spelled disaster for the iron works.[2] It started in the engine house while the forge was in full blast and, aided by a strong wind, it quickly spread and destroyed the forge, machine shop, engine house, and coal house containing 20,000 bushels of charcoal. Except for 10,000 bushels of charcoal that were saved by covering with sand, the iron works was a complete loss.[3] This was the end of the Marquette forge as the Cleveland Iron Mining Company never rebuilt it. The forge produced about 800 tons of iron while in existence.[4]

Watt Eaton had charge of the Carp River Forge at this time and was also turning out a considerable amount of iron. Ben Eaton had left the country in the spring of 1851, financially ruined.

The combined total of iron blooms shipped by the Sharon and Cleveland Iron companies for the 1853 season was 900 tons, with 500 tons of iron ore. The Cleveland Iron Mining Company, with the loss of the forge, put nearly all of its crew mining and sledding ore down from the mines in January of 1854, in preparation for the coming shipping season.

As the quality and quantity of Lake Superior ore became known, demand for it at the eastern and lower lakes furnaces increased greatly, but the poor roads hindered its delivery from mines to port at Marquette. In response to the rising demand, Heman B. Ely of Cleveland, and other engineers, left Detroit for Marquette in May of 1852 to survey the 15-mile route from the interior mines to the lake shore for a railroad.[5] Clearing

---

[1]*Lake Superior Journal*, November 12, 1853.

[2]*Mining Journal*, January 29, 1876.

[3]*Lake Superior Journal*, July 12, 1855, noted ". . . The flames had wrapped themselves around the ill-fated works and left nothing but a heap of blackened ruins and some chimneys to tell where it once stood."

[4]*Ibid.*

[5]*Lake Superior Journal*, May 22, 1852.

and grading was completed on the first five miles of the line in the fall of 1853.

Even though the winter snow roads from the iron mines to Iron Bay at Marquette were said to be ". . . equal to any plank road for smoothness,"[6] the Sharon Iron Company projected a plank road from the Jackson Mountain mine to the lake in 1853, and by May of 1854 over half of this distance was graded. The Cleveland Iron Mining Company then surveyed an extension on the plank road early in 1854, to run from Jackson Mountain to its mines three miles further west.[7]

During the construction of the plank road and the Lake Superior Iron Company's railroad, many different names were applied to both undertakings: Iron Mountain Rail Road, Iron Mountain and Lake Superior Rail Road, Carp River and Iron Bay Plank Road, Lake Superior Iron Company Railroad, Iron Mountain Railway, and the Marquette and Bay de Noc Railroad. It seemed that names changed seasonally.

Eagle Mills location was originally settled by Duncan and Parrish. Duncan and Parrish built a steam saw mill there while the plank road was under construction and did a flourishing business. The mill, employing about 35 men, could cut up to 15,000 feet of plank per day. They had 900,000 feet of three-inch plank cut for the Sharon Company road by May of 1854, enough for half the distance, and had a contract to cut and lay all the plank needed.[8] The road was opened to traffic on November 1, 1855, with mule- and horse-drawn single cars carrying about four tons of iron ore per trip, making one trip each day. Later, strap iron rails were added to the plank road which aided the teamsters in keeping the loads of ore on the road.

Four miles upstream from Lake Superior on the Dead River, William G. Butler and Matthew McConnell began construction of the Buckeye Forge in the fall of 1853, where iron making would continue for nearly a quarter of a century. During the first winter, ore was sledded down from the Jackson Mountain, as the Buckeye Company had no mine of its own. Work was also started on a water-powered sawmill. Charcoal was readied for the forge and in July of 1855, it started making about 10 tons of iron blooms a week.

On September 22, 1855, articles of association were filed by Butler and McConnell, William S. McComber, A. J. Bennington, Peter White (the president), M. H. Maynard, W. J. Gorden, M. L. Hewitt and J. G. Butler, changing the Buckeye Forge and Company into a stock company titled the Forest Iron Company capitalized at $25,000 with the sale of 1,000 shares at $25 each.

The Collins Iron Company, builders of the fourth and last forge in the Upper Peninsula, was organized on November 8, 1853, with a capital stock

---

[6]*Lake Superior Journal*, November 10, 1852.

[7]*Ibid.*, June 3, 1854.

[8]*Ibid.*, June 3, 1854.

of $500,000. The incorporators were Edward K. Collins (who had the controlling interest in the business for the purpose of obtaining iron to make boiler plate and shafts for his steamship line), Elon Farnsworth, Edwin H. Thompson, Robert J. Graveraet, and Charles A. Trowbridge. This concern was located two miles below the Buckeye Forge on Dead River.[9]

The earliest references concerning the Collins Company are in the February 1, 1854, issue of the *Lake Superior Journal*, which quoted a letter dated January 27, 1854: ". . . The Collins Iron Company, is cutting cord wood - clearing land - getting in saw logs on Dead River preparatory to building in the spring."

It was necessary to blast in the rock bottom of the river to make room for the races and water wheels to obtain good water power. The company owned a large tract of hardwood and pine adjacent to the river site and held title to some valuable iron lands near Jackson Mountain. The steamer *General Taylor* delivered an anvil block weighing 11,640 pounds at Marquette in June of 1855, destined for use under the heavy triphammer at the Collins Forge. The company began making bloom iron at the forge in August of 1855, producing eight tons of bar iron per day.

The residents of the Upper Peninsula recognized at an early date the need for a canal with locks at Sault Sainte Marie to improve commerce in the Lake Superior country. The shipping of iron, iron ore, lumber, copper and fish to the markets of the east and the lower lakes increased each season. At the same time the growing population of the Peninsula created a demand for more consumer goods from the lower lakes. The high cost of portaging at Sault Sainte Marie caused merchandise to be priced beyond the means of many residents. Clamor for a canal went on for many years until the State Legislature finally answered the need and authorized its construction. The canal opened for business on June 18, 1855, and on August 17 the southbound brig *Columbia* carried the first load of 100 tons of iron ore through the locks.

At Carp River, Watt Eaton continued to operate the forge until the fall of 1853 when it was abandoned. A group of forgemen from Clinton, New York, with Azel Lathrop formed the Clinton Iron Company and leased the forge in September of 1854, operating it until May of 1855 when it was shut down. A newspaper reported, "It seems to have 'busted' them as it has "busted" everyone who had anything to do with it.[10] The company's real problem had been a lack of operating capital.

Following the Clinton Iron Company, Peter White leased the Carp River Forge but soon gave up on it and "surrendered the lease to the owners of the plant." This attempt was followed by that of J. P. Pendill, who "retired soon after taking hold, the undertaking failing to prove profitable even under his energy."[11] This last short venture by Pendill was the end

[9]Where the present bridge crosses the river at the west end of Wright Street.
[10]*Mining Journal* (Weekly Edition), January 16, 1892.
[11]*Proceedings of the Lake Superior Mining Institute* (Ishpeming, Mich., 1914), XIX, p. 301.

of the Carp River Forge and it was abandoned.[12] Though a financial failure most of its life, it was of major significance in the development of the iron industry in the Upper Peninsula.

During the first week of October, 1854, a Lake Superior gale devastated Marquette Harbor. The massive timber pier built that season for the Sharon Iron Company was battered to pieces. Of the $15,000 in labor and materials put into it, all that remained after the storm were timbers strewn on the sandy beach. The loss of the pier was a great setback for the company and the growing settlement at Marquette. It would take almost a full year before the pier would be replaced and the harbor considered safe. High winds of the gale knocked down virgin timber all along the routes of the plank road and railroad then under construction, blocking them firmly and considerably slowing work.

COLLINSVILLE

Power lines, roads and water flumes for a hydroelectric station have obliterated all traces of this forgotten hamlet where many industrious and resourceful people found employment under Graveraet, superintendent of the Collins Iron Company. In the fall of 1855, the "horrid" smell of the burning charcoal pits was everywhere in the community, which also grew accustomed to the deafening sound of the huge trip hammers, breaking up the ore fresh from the roasting kilns and hammering the red hot blooms just lifted out of the fires. There was an abundance of water to turn the waterwheel, and a fall of 26 feet supplied the blowers and hammers with all the power needed. In addition to the forge were the stamp house, blacksmith shop, bellows house, and kiln house, all in operation. The wood choppers had cut large clearings for the houses of the residents and the company store.[13]

The Collins Iron Company had a contract with the Collins Steamship Company to supply 400 tons of iron by the close of navigation in 1855, and this contract had been fulfilled by the end of November.

The *Lake Superior Journal*, on February 2, 1856, listed all expenses and materials that went into making iron in January of that year, and concluded that a $39 a ton profit was being made in the operation of Collinsville:

---

[12]The June 20, 1857, issue of the *Lake Superior Journal* stated, ". . .there are three forges in this vicinity, combining 12 stacks for fires that manufacture bloom iron of the best quality, worth $65.00 per ton here . . . ." The Carp, Forest, and Collins were the only forges left as the Marquette Forge burned in 1853.

[13]*Lake Superior Journal*, December 1, 1855. (The newspaper was moved from Sault Ste. Marie to Marquette in 1855 and later became the *Mining Journal*.)

Four fires were used during the month and from each fire, in
a 24-hour period, 2,400 lbs. were made, for a total of 9,600
lbs. a day, making 112 tons in the month, worth $80 per ton $8,960

This required 26,000 bushels of charcoal costing $70 per thou-
sand (Graveraet had the contract to supply this)   1,820

The purchase, cartage, roasting, stamping and delivery to the
forge of 224 tons of ore at $7 per ton   1,568

Average wages of 12 workmen ($2.83 per day)   883

Two assistants   50

One Smith   50

Repairs, $1 per ton   112

Incidental expenses   112

                  Total Cost $4,595

The Collins forge was shut down for the winter 1856-57 because of the
loss of materials in the burning of the steamer *B. L. Webb*. The steamer
left the Soo in late November of 1856 bound for Marquette. It was loaded
with the iron company's winter supplies and iron goods for the railroad
then under construction. Nearing Waiska Bay, rough seas and heavy snow
were encountered forcing the vessel to anchor in the bay to wait better
lake conditions. Here, somehow, the *Webb* caught fire and burned to the
water's edge, destroying the valuable cargo.

THE FOREST IRON COMPANY

The fires were shut down at the Forest Iron Company in November of
1855 to allow the workmen to restock ore and charcoal supplies, and in
February of 1856, the two fires were again in operation. The location of
the iron works was between the river and a ridge of high bluffs, where
enough flat land was available for buildings and farming.[14]

Farming was carried on extensively on the land cleared by the wood
choppers surrounding the location, and in 1856 a fine crop of potatoes,
rutabagas, and oats, cut green for fodder, was harvested. The buildings
at the iron works were rather crude but comfortable; there were seven
dwelling houses with barns and other out-buildings, the forge house with
stamping and bellows rooms under the same roof, and the saw mill office
and saw mill.[15]

The water powered saw mill was capable of cutting 3,000 feet of lumber
in a 24-hour period with only two men working it, and a saw mill operated
at this site for many years. The river at this point offered many advantages for water power and without too much difficulty a dam and flumes
were built to divert the water, producing the 12-foot waterfall that was
necessary to apply power.

---

[14]This site is presently reached on the road leading to the McClure hydroelectric plant, and lies between the steel
bridge and the public fishing site.

[15]*Lake Superior Journal*, February 2, 1856.

BEECHER FURNACE

AND

Marquette Rolling Mill

Marquette, Mich.

Miners of Iron Ore

And Manufacturers of

Pig Metal, Muck Iron,

AND

Merchant Bar Iron.

Orders from the trade respectfully solicited.

W. L. WETMORE, President.

PETER WHITE, Secretary.

W. W. WHEATON, Treasurer and General Agent.

CHARLES JENKINS, Assistant Agent.

From *Beard's Directory of Marquette County*, 1873.

Chapter III

THE ERA OF THE BLAST FURNACE IN THE UPPER PENINSULA

The Pioneer Iron Company was organized by Charles T. Harvey of Marquette on July 20, 1857. Harvey and his partners — Moses A. Happock of New York and William Pearsall of Jersey City, New Jersey — filed articles of association which allowed an initial capitalization of $125,000 through the sale of 5,000 shares of stock at $25 each.

Prior to the formal organization of the firm, the trio had contracted Stephen R. Gay and Lorenzo D. Harvey of West Stockbridge, Massachusetts, to construct the company's furnace and stacks where Negaunee is now located. Gay and Harvey arrived at Marquette abroard the steamer *General Taylor* on July 3, 1857, accompanied by 22 workmen and their families. Crude shelters were constructed near the site and then work was begun on the first stack of the Pioneer Furnace. Machinery for the furnace was ordered from a firm located in Cold Springs, New York, with the exception of two steam boilers which were salvaged from the old Marquette Forge and transported to the site over the plank road.

The new steam railroad was completed from Marquette to the vicinity of the Pioneer Furnace in August of 1857 and the first materials transported by the railroad were five car loads of brick for the new furnace. With the completion of the railroad to the Jackson Mine in September, the need for the old plank road and strap iron railway came to an abrupt end.[1]

During the construction of the Pioneer, Gay leased the old Collinsville forge in Marquette for his own use. He started converting the forge into an experimental blast furnace on January 18, 1858, in an attempt to see if good pig iron could be made in a small furnace. The furnace was completed on January 20. Gay began his experiment on Januay 21, filling the stack of the furnace with charcoal, ore, and flux in proper proportions. It ended four days later after producing 5,008 pounds of pig iron — the first made in the Upper Peninsula.

Construction on the Pioneer Furnace was completed in February, 1858, and in April the boilers were fired to test the steam engine and piping. The first pig iron was produced there on April 26, 1858, when Lorenzo Harvey tapped the furnace.[2] The fuel used in this first run was pit charcoal made by Antoine Barabee in the vicinity of present-day Iron Street in Negaunee. He arrived there in October of 1853, and, with his brother Joseph, cleared the timber from the land where the Pioneer Furnace was erected in 1857-58.

[1] Williams, Ralph D., *The Honorable Peter White* (Cleveland, 1907), p. 146.
[2] The Pioneer was the first full-sized blast furnace to produce pig iron in the Peninsula.

THE COLLINS BLAST FURNACE

Work was started on the foundation of a full-size blast furnace at Collinsville on August 2, 1858, under the direction of Stephen Gay. The furnace was built at the foot of a high bluff with the top of the stack level with the ground on which the storage houses for the firm's charcoal were built. Kilns for charcoal and ore were built adjacent to the furnace at the foot of another bluff. The kilns were filled through an opening on the top, and both types had doors at the bottom sides through which they could be emptied of charcoal or furnace-ready ore. The site for this construction was near Gay's experimental furnace. A firm headed by Mr. Dodge, who had at his command "any amount of capital he desired," leased the new furnace for the high rent of $8,000 annually.

After a short construction period, the furnace was put into blast on December 13, 1858, "vomiting forth its molten, seething mass of iron."[3] The furnace was turning out eight tons of pig iron by its fifth day of operation, using only 90 bushels of charcoal to the ton of iron. Its maximum production was expected to reach an output of 12 tons per day.

The Collins Furnace was still producing pig iron in 1862 with an average production of nine tons per day. At the time, 50 men were employed at the works by the leasee, Stephen R. Gay, who had turned the management of the works over to Timothy T. Hurley.

Collins blast furnace and casting house, Collins-ville, 1888.

MARQUETTE COUNTY HISTORICAL SOCIETY

PIONEER NO. 2 AND THE NORTHERN AT HARVEY IN BLAST

The Pioneer No. 2 Furnace in Negaunee was the third blast furnace built in the Upper Peninsula and started making an excellent grade of charcoal iron in May of 1859. Pioneer No. 1 stack was blown out on January 29,

[3]*Lake Superior Journal*, January 5, 1859.

1860, for a new brick lining after making 6,688 tons of pig iron in three blasts. This iron was produced at a cost of $24 a ton but sold for only $22. This financial setback discouraged the initial investors who leased the furnace to I.B. Case in 1860 for a four-year period. Case turned the furnace into a profitable concern and employed 60 men at the furnace by 1862, while an additional 10 men were needed to perform other work.

The Pioneer Furnace suffered its first major fire on August 9, 1864, when the No. 2 stack was destroyed. Repairing the damage took nearly six months and on January 15, 1865, it was again making iron.

Stone stacks of the Pioneer Iron Company, Negaunee.

MARQUETTE COUNTY HISTORICAL SOCIETY

On September 15, 1864, the Iron Cliffs Company was formed, having a capital stock of $1,000,000 in 40,000 shares at $25 each. Those having interests in the company were William B. Ogden and John W. Foster of Chicago, and Samuel J. Tilden of New York. In the spring of 1866, the Iron Cliffs Company gained a lease of all the Pioneer Iron Company property, which included the two furnaces, 4,000 acres of land and part of the village of Negaunee, also an ore lease from the Jackson Company. Eventually, the Iron Cliffs Company obtained ownership of all the property.

Charles T. Harvey organized yet another company during the winter of 1859 with the aid of John C. Tucker and Moses A. Hoppock of New York. Articles of Association were filed by the trio on May 16, 1859, for the formation of the Northern Michigan Iron Company. An initial capitalization of the company was accomplished by the sale of 5,000 shares of stock at $25 per share.

The location of the new company's furnace was four miles south of Marquette at the mouth of the Chocolay River. The furnace was built under the supervision of Lorenzo D. Harvey and was originally designed for fueling with anthracite coal. It went into blast during the summer of 1860 but produced metal for only two months when the owners were forced to shut it down due to the exhaustion of their coal supply. Settlers in the Harvey area were clearing acres of land of hardwood to establish farms and suggested to the ironmakers that they convert the furnace for charcoal fueling. They convinced the company, and in 1861, the furnace was rebuilt to operate on charcoal. The farmers supplied pit charcoal to the firm at seven dollars per hundred bushels, until the company built one sheet iron, brick-lined charcoal kiln and ten square brick charcoal kilns.

Northern Furnace, Harvey, in blast about 1891.

After producing an average of 14 tons of pig iron a day, the Northern furnace blew out on December 5, 1862, for repairs. They had on hand 50,000 bushels of charcoal and employed from 80 to 100 men. While the fires were down, nine additional iron charcoal kilns were built making a total of ten iron and ten square brick kilns. They could turn out 2,000 bushels per day and, with iron selling at $40 per ton, the company wanted the furnace to have a long, successful run.

During this same period, Stephen Gay also supervised construction of a furnace in the Marquette area. This furnace, which was called the Bancroft, was located at the site of the old Forestville Forge in order to take advantage of the existing water power. The furnace went into production in May, 1861. A year later it had turned out more than 3,500 tons of pig iron, "in addition to which, in the forge . . . has been manufactured a considerable quantity of superior bar iron."[4] The Forestville smelting works was prospering in May of 1862 and its furnace, the Bancroft, was considered the finest in the Upper Peninsula. All the forests about the village had disappeared and a plank road was under construction from Gay's siding on the Marquette, Houghton & Ontonagon Railroad to the furnaces at Collinsville and Forestville to ease the handling of ore and pig iron. Seventy-five men were employed at the Bancroft by November of 1862.

---

[4]*Lake Superior Journal*, May 22, 1862. (This is the last reference found of the old forge being in operation and presumably it was abandoned after this date.)

With the Civil War in progress, the demand and price of pig iron was such that the four iron companies in Marquette County needed all the help they could get to make charcoal. The *Lake Superior Journal* ran this ad on February 20, 1863:

WOOD CHOPPERS WANTED! I want 20 wood choppers — will pay 75 cts. per cord cash. Parties from abroad can rely upon employment and their money. I.B.B. Case, Pioneer Iron Works, Feb. 11, 1863.

The following week a similar ad appeared in the *Journal:*

ATTENTION: Wood Choppers! 80 Cents a Cord. The Northern Iron Co. will pay eighty cents a cord for all wood cut before May 1st. Inquire of L. D. Harvey, Harvey, Feb. 27, 1863.

This was another furnace in which Peter White had an interest. On September 12, 1864, the Bancroft Iron Company filed articles of association, and under the charter allowed a capital stock of $250,000 in 10,000 shares. Corporators in the firm were Samuel L. Mather, John Outhwaite and William L. Cutter of Cleveland, William E. Dodge of New York, Peter White and Samuel P. Ely of Marquette, and Henry L. Fisher and L. S. McKnight of Detroit.

Chapter IV

THE MORGAN IRON COMPANY

Cornelius Donkersley started preparations for construction of an iron works on the Little Carp River (Morgan Creek) eight and one-half miles west of Marquette at the same site where a Mr. Schweitzer had operated a grist mill in 1856. Donkersley first cleared the land and built houses for the workmen and then began removing rocks from the site where George Craig would build the huge stack for the Morgan Company. George Rublein had received the contract to build and operate 21 charcoal kilns of 40-cord capacity for the company. These charcoal kilns were expected to produce a daily average of 1,700 bushels. By May 8, 1863, more than a dozen houses had been built for the workmen and supervisors and many more sites were laid out for future development on a plateau overlooking the creek.

The initial work underway, Lewis Henry Morgan of New York, in partnership with Samuel Ely of Marquette, established the Morgan Iron Company with the signing of articles of association on July 1, 1863. The company had an initial capitalization of $50,000 furnished by the sale of 2,000 shares of stock at $25 a share. Morgan was also an investor in the early railroads of Marquette County and served on the board of directors of the Iron Mountain Railroad. Other investors in the company were Joseph S. Fay, Harriet Ely, Ellen White, and Cornelius Donkersley, who served

Village of Morgan in July, 1864.

as Company Agent, Iron Master, and Superintendent of both the construction and the operation of the furnace.

The company also purchased 1,600 acres of choice maple lands located in the vicinity of the furnace to secure enough wood to supply their kilns for several years. The Morgan Furnace was located at the foot of a bluff which was only 450 feet from the Marquette & Ontonagon Railroad. Marble taken from the stack's construction site served both as building material and, for a time, as flux once the furnace was in operation.

The furnace was lit on November 27, 1863, and by December 31 had already produced 337 tons of cast pig iron. In its first ten months of operation, it netted a profit of 220 percent which allowed the company to pay its indebtedness for land and machinery and declare a 100 percent dividend for the investors.

About one mile east of Lake Michigamme, near the western terminus of the Marquette & Ontonagon Railroad, the Morgan Iron Company began construction of a blast furnace in the spring of 1867 — later named the Champion. The stack measured 20 feet square at the top and bottom, had a nine-foot bosh, and was built on the side of a high bluff facing a swamp. In late July, 1867, the stack was completed but the huge pieces of machinery, made at the Washington Iron Works in Newbury, New York, still needed to be placed in position.

Water for the furnace was piped from a nearby stream running from the hills down to a large storage tank near the "top house," assuring plenty of pressure for the iron works lying below. One large steam engine supplied the power for blowing and hoisting, and charcoal would be charged from the top house. Six kilns out of a planned 12 were smoking in July, but John R. Case, the superintendent, wanted to be sure of a good charcoal supply when the furnace went into blast, so he also had the colliers at work making pit charcoal.

The "Champion" furnace went into blast on December 4, 1867, and in a matter of weeks proved to be a very capable producer. From the beginning the furnace was using only 110 bushels of charcoal to the ton of iron. Founder James Dundon achieved a charcoal-to-ton ratio of only 79 bushels in April 1869. By October, 1869, the Champion was making from 128 to 171 tons of pig iron a week which was considerably more than the average output of other local furnaces.

GREENWOOD FURNACE OF THE
MARQUETTE & ONTONAGON RAILROAD

The Marquette & Ontonagon Railroad Company, which operated a line from Marquette to Michigamme, entered the iron business in 1865 with the construction of Greenwood Furnace 12 miles west of Negaunee. A small community developed at the site. It consisted of about 50 frame houses, a sawmill, a company store and the furnace. The furnace was placed in blast during June of 1865 and had a successful run through December of 1867, producing 5,339 tons of pig iron during that time. The furnace blew out for major repairs in January, 1868, with the stack almost completely dismantled by February. The furnace was rebuilt with hand-quarried stone which the furnace laborers cut from a quarry on the south side of the railroad. While the work was in progress, the Greenwood Furnace and 8,000 acres of surrounding timber lands were purchased from the railroad company by the Michigan Iron Company and the repaired furnace was again in blast on August 21, 1868.

Michigan Iron was capitalized at $150,000 and its first purchase had been 6,000 acres of hardwood lands on the M & O Railroad. President of the company was A. B. Meeker of Chicago; vice-president, H. J. Colwell of Clarksburg, and the secretary-treasurer was A. G. Clark of Marquette.

The company chose Donkersley's abandoned sawmill at Clarksburg as the site for its new furnace and on June 3, 1866, construction was started. The site soon grew into a small iron and lumber producing community and had a post office, general store, a drugstore, a blacksmith shop, a Catholic church and nearly 50 houses for the French-Canadian workmen.

The Michigan Furnace was equipped with nearly 100 tons of massive machinery. The blast producing engine was described as consisting of "
. . . two cast iron, double acting vertical engines, each 45 inches in diameter, with five-foot stroke worked by an engine of 70 inches diameter of cylinder, and four-foot stroke."[1] Sweeney & Company of Wheeling, West Virginia, manufactured the furnace's hot blast engine, while the Marquette & Ontonagon Railroad shops in Marquette manufactured the 15-horsepower engine that operated the Blakes ore crusher.

Fifteen kilns were built at first to supply charcoal for the furnace and had produced 110,000 bushels of charcoal by the time the furnace went into blast on February 10, 1867. To meet ever increasing fuel consumption, the company erected five more charcoal kilns during 1867. Pig iron was cast regularly at the Michigan furnace until November, 1867, when the furnace was blown out for replacement of the hearth. Production from February until November had been 3,900 tons.

The stone and masonry work on the new hearth was completed on December 1, 1867, and soon after the furnace was back in blast. This second run was prematurely terminated when leaks were discovered in the hot blast pipes. The furnace had to be shoveled out so the necessary repairs could be accomplished. Repairs were finished by January 1, 1868, and the furnace was soon in blast again. Total production for the Michigan Furnace in 1867 was over 4,100 tons of pig iron.

---

[1]*Lake Superior Mining & Manufacturing News*, Negaunee, June 8, 1867.

The Michigan Furnace, Clarksburg, had been long idle when this photo was taken around the turn of the century. From *Lake Superior Mining Institute* proceedings, 1903.

SCHOOLCRAFT IRON COMPANY AT MUNISING

Peter White and H. R. Mather formed the Schoolcraft Iron Company at Munising in June of 1866 from the skeletal remains of three earlier companies. The first of these had been the Munising Company, founded in 1850 by a group of Philadelphia men for the purpose of developing a resort community on the southeast shore of Munising Bay. A townsite was planned for Munising in an area which had long been inhabited by Ojibwa Indians, and a resort hotel and other developments were envisioned. However, with its capital soon depleted the company ceased operations and, in 1856, it was taken over by the newly-formed Grand Island Iron Company.

The Grand Island venture planned construction of an iron furnace and sawmill and laid out a "paper plat" of Grand Island City. A few lots were sold, part of a dock built, and a hotel and several dwellings were erected. A road was cut through to Bay de Noc on Lake Michigan following an old Indian trail. The Grand Island Iron Company also lacked adequate capital reserves. It continued in operation but a short time and went out of business. Settlers remained at Munising and survived by living off the land.[2]

The third company which furnished a foundation for the Schoolcraft Iron Company was founded in 1863 by S. F. Church and H. R. Mather as the Grant Mineral Land Company. Grant Mineral Land Company purchased lands in Marquette, Houghton, and Menominee Counties and in 1863 increased its holdings with the purchase of 40,000 acres of hardwoods in Schoolcraft (now in Alger) County. Financial setbacks forced a reorganization of the company in 1866 as the Grant Mining and Manufacturing Company.

---

[2]*Mining Jounal*, November 21, 1868.

White and Mather purchased the holdings of these two companies and organized Schoolcraft Iron Company "under the General Mining laws of the State of Michigan with a nominal capital of $500,000."[3]

The site for the Schoolcraft Furnace was on the south shore of Grand Island Bay where East Munising is now located. Actual construction started on the furnace on May 17, 1867. The foundation for the stack was blasted from soft rock on the bank of Munising Creek, 1,100 feet from the shore of Lake Superior. Stone for the furnace stack was quarried by the workmen on Grand Island and floated across the bay on scows to the building site, which increased the final cost of the structure.

Mechanical components for the Schoolcraft Furnace were made at the Washington Iron Works, Newburg, New York. One large engine furnished the hot air blast for the furnace, while two smaller engines were purchased to operate the ore crusher and the water pump. An 8,000 gallon water tank was built above the works on a hill and filled by diverting water from Munising Creek. The tank was intended as a safety measure in case the furnace caught fire.

The furnace was equipped with a water balance that raised the crushed ore and limestone to the stack top. Five kilns were constructed at the furnace location and six additional kilns were constructed four miles out into the hardwood forests to produce the charcoal needed for the furnace. An additional 150 charcoal pits were constructed to augment the production of the charcoal kilns.

Start up of the furnace was delayed due to the late arrival of necessary castings and hearth stones, with the furnace finally going into blast June 28, 1868, after the needed furnace parts had been received and set in place. Initial output of pig iron was low due to water-logged charcoal, but it later increased to 100 tons per week.

A cribbed dock over 600 feet in length and 80 feet in width was built to accommodate the ships bringing ore and other necessities to the furnace and the developing settlement. In the fall of 1868 the Munising location was prospering and the workmen were living fairly well. William A. Cox ran a well-stocked general store and a doctor had located his office in the community. Lewis Williams was the company agent, William Shea the foundryman, and J. T. McCollum was the company bookkeeper.

The Schoolcraft Company did fairly well for about two years, but in October, 1870, it was discovered that the company was hopelessly mired in debt.[4] Management officials in Philadelphia had overextended the company financially and many months' wages were owed to the workers. The supply of ore on hand at the furnace ran out forcing it to shut down. Day-to-day living conditions at Munising became so bad that on November 12, 1870, most of the employees and their families left for Marquette on the mail boat, while a few hardy individuals remained behind to tend

[3]*Mining Journal*, Novmber 21, 1868.
[4]*Ibid.*, October 29, 1870.

Schoolcraft Furnace, Munising, in blast, c. 1870. Two men (right) are standing near a "charging buggy."

the charcoal kilns.[5] Peter White purchased the bankrupt company for $65,000 on July 12, 1871, and organized a new company to operate the furnace at Munising.

The Schoolcraft Furnace was put in blast again on April 8, 1872, with a stock of coal and ore on hand sufficient to last until the opening of navigation that year, and remained in blast until April, 1873, when it was blown out for a lack of coal. For the year 1872 the furnace had produced 2,500 tons of pig iron. The Schoolcraft (renamed the Munising) was blown out in the fall of 1873 after a short blast so that a bell and hopper could be added to the stack. This apparatus allowed charging of the furnace without any loss of gas. Six new kilns were built at the Munising during the same period.

When the furnace was placed back in blast it was charged with a mixture of 20 percent soft hematite and 80 percent hard specular ore. Using this mixture the Munising was producing 31 tons per day in March, 1874, utilizing only 107 bushels of charcoal to the ton. The furnace ended a 24 week blast on September 7, 1875, when it ran out of both ore and fuel, but during this time it had produced 4,634 tons of pig iron, reaching a one day high of 37½ tons. Only nine of the original 21 furnaces on the Upper Peninsula were in blast in September, 1875. The Munising stayed down until sometime in the summer of 1876 and was then leased by Major Henry Pickands of Onota, also called Bay Furnace, which was located on the west side of Grand Island Harbor. He ran the Munising Furnace until November, 1876. It was next leased by Dan Rankin who arranged for the tug *J. C. Morse* to make four trips a week from Marquette to deliver ore to the furnace. Rankin placed the furnace in blast on July 14, 1877,

[5]*Mining Journal*, November 19, 1870.

and operated it until November 24 when it was blown out due to lack of fuel and ore supply. Plans were to idle the furnace only for the winter, but the Munising was never warmed again. Final disposition of the furnace took place in 1901 when the Lake Superior Iron and Metal Company of Hancock purchased all of the old machinery and shipped it to Hancock.

MARQUETTE & PACIFIC ROLLING MILL

Peter White's financial interests in the iron industry were widespread. Only four months after founding the Schoolcraft Iron Company, White helped organize the Marquette & Pacific Rolling Mill Company. Articles of association were filed on October 1, 1866, with an initial capitalization of $500,000 in 20,000 shares of stock at $25 each. Besides White, other investors were John and William Burt, Samuel P. Ely, Edward Breitung, Timothy T. Hurley, Cornelious Donkersley, Alvin Burt, and W. L. Wetmore, all of Marquette.

The furnace envisioned by White and his fellow investors was to be unique in the Upper Peninsula. Other furnaces to this date were fueled with charcoal (except for the Northern's brief use of anthracite coal) and had only produced cast pig iron. The planned furnace was fueled with bituminous coal and, in its rolling mill department, puddling and heating furnaces and trains of rolls would turn out muck bar and merchant bar iron. The charcoal furnaces built in the Upper Peninsula reached heights slightly over 40 feet, having boshes of 9½ feet with bases of about 20 feet square but the Marquette & Pacific was designed to be 60 feet high with boshes of 15 feet, and a base measuring 40 feet square.

Construction started on this mammoth stack in mid-1867, and its location was near the present Shiras Steam Plant. In September of 1867, the stack was raised above the arches, and in the rolling mill a double puddling furnace and a heating furnace were about completed, with one steam engine in to run one train of rolls. In January of 1868, the stack was raised 40 feet, but work was suspended for the winter when the stack settled nine inches on the lake side.[6]

On September 1, 1868, the rolling mill was put into operation, having an initial capacity of five tons of bar iron per day, using one heating and one puddling furnace. By March of 1869 the stack was up to a height of 52 feet, and the rolling mill had turned out 594,795 pounds of finished iron since starting. The pig iron used in the mill was purchased from the other iron works operating in the county.

The Marquette & Pacific Furnace was completed in March of 1870 and in May, lake vessels discharged coal, coke, and limestone on their dock in preparation for the first blast. In July of that year iron was selling in Pittsburgh at a depressed price of from $30 to $32 a ton, and in a letter

---

[6]*Lake Superior Mining & Manufacturing News*, January 2, 1868.

to the editor of *The Mining Journal* on July 30, 1870, one of the Burts stated that". . . Our furnace and mill had much better lie idle for another year or longer, if need be, than to go into blast with such a market before us . . . ." And, true to his word, the furnace did lie idle for a year.

The rolling mill went into production in June of 1871 making iron for the breakwater then under construction and on July 13, fire was finally lit in the stack and the blast was turned on. The furnace was tapped on the evening of July 14 and the first metal was run out. John Stevens of Sharpsville, Pennsylvania, served as the furnace founder and had many years experience with bituminous coal furnaces in Pennsylvania. Under his care the furnace was soon making 22 tons of iron a day. The furnace was charged with a mixture of hematite and slate ore from the Lake Superior Mine in Ishpeming and coal of the Briar Hill type from Cleveland, Ohio. Daily production continued at the furnace until October 7, 1871, when the furnace was blown out for repairs.

The Marquette and Pacific Furnace went back into blast after the hot blast had been moved from the top of the stack to the ground in January, 1872, and production resumed at the rate of 32 tons a day through June when, as a local paper reported, the furnace had ". . . just recovered from a three weeks wrestle with the 'chills' caused by the burning of a tuyere. Her recovery is due to the skill of Dr. Stevenson (the founder), who applied the necessary stimulants, and by frequent bleeding over the dam, rescued his patient from a threatened salamander."[7]

The Marquette & Pacific production for 1872 was 4,332 tons of pig iron and 622 tons of muck bar from the rolling mill.

A new board of directors were installed for the Marquette & Pacific Furnace company in March of 1873. William W. Wheaton, ex-mayor of Detroit, was placed in charge and work was immediately started to rebuild the stack. By July the original stone stack was entirely dismantled and a new iron shell stack 60 feet high with boshes of 15 feet were erected in its place. Two new Player hot blast ovens were also installed. Over the years a total of $247,000 had been invested in the iron works.

The company owned 24 acres on which the iron works was situated with a frontage of 1,400 feet on Lake Superior. A new dock extended 500 feet into the bay for the receiving of fuel, flux and machinery and it was also used for shipping of iron. The water on either side of the dock was deep enough to accommodate the largest ships on the lakes. (In September of 1873, a bad storm hit and the schooners *E. C. Roberts* and *A. H. Moss*, both unloading coal at the dock, were scuttled by their masters and sunk to protect them from being broken up in the heavy seas — only their masts stood above the water.)

The furnace went into blast on July 24, 1873, with the first iron run out on the 28th, and it was said that ". . . this concern so long a

---

[7]*Mining Journal*, June 29, 1872.

Marquette & Pacific Rolling Mill Furnace.
From *Lake Superior Mining Institute*, proceedings, 1903.

disgrace to the city, may at last be a success . . . . ."[8] In a three weeks run 25 tons of No. 1 bituminous foundry iron were run out daily and the founder, John Fisher, expected a higher output when it was thoroughly warmed up. The rolling mill averaged nine tons of muck bar iron per day, with a high of 13 tons in one day, and some bolt rod was turned out on the merchant bar rollers. The mill was set up with the muck and merchant bar trains of rolls, one heating furnace and eight puddling furnaces. After only 26 days of casting when 710 tons of iron were made, the furnace was blown out with a badly worn lining. A "conflagration" was thought by some to be the only answer for this badly-built and badly-managed iron works.

Like the Menominee, the Marquette & Pacific went through a name change. Luther Beecher of Detroit was heavily invested in the Marquette & Pacific furnace and owned the Rolling Mill Mine located in Negaunee. His son, George L. Beecher, was supervisor of the dock and furnace grounds in Marquette. In August of 1873 the board of directors decided to change the name of the furnace to the "Beecher." Luther Beecher was well known, owner of banks in New York and Detroit, and more than likely the furnace was named after him.

At this time newspapers all over the country were reporting the Beecher-Tilton scandal. The main characters were famous clergyman Henry Ward Beecher, Elizabeth Tilton and her husband Theodore.

In 1855, Beecher married Elizabeth and Theodore Tilton (Tilton was editor of the Brooklyn *Union* and the *Independent*) and for many years this threesome were close friends. Eventually, Mrs. Tilton confessed to her husband of her infidelity with Beecher, which eventually led to Tilton swearing out a complaint against Beecher in August of 1874, charging him with alienating the affections of Mrs. Tilton and asking $100,000 for

---

[8]*Mining Journal*, August 2, 1873.

damages. The case came to trial in January of 1875 in New York and, after six months of hearing evidence, it ended with the jury voting nine to three in favor of the celebrated preacher.

In October of 1874, less than two months after Tilton filed charges, it was reported "Since the Brooklyn affair . . . the Marquette & Pacific Rolling Mill furnace has lost her distinctive name 'Beecher.' The board of directors concluded to try their luck of another name,"[9] and the name Marquette & Pacific was again put to use.

The furnace was repaired and in late September it was blown in on bituminous coal and coke, making at times 40 tons of iron a day. In January of 1875 the company purchased all the coal on the Northern Furnace dock at Harvey, which was enough to keep it in blast until late May.[10] The puddlers in the Marquette & Pacific rolling mill went on strike in January because of an impending reduction in wages, and the rolls were stopped permanently.

The furnace made a high of 52 tons in March of 1875, and continued to make 40 to 45 tons daily, and for the month of July, 1,129 tons of Bessemer iron were cast. In March of 1877, the M & P was blown out after making 25,452 tons of iron on one hearth, in a run of 27 months; it was banked for two months, waiting for a shipload of coal. Frank Skelding, one of the better founders in the United States, was in charge and resumed making iron in May.

E. C. married 1857 & took bride to Marquette Mich.

Hungerford

DEER LAKE FURNACE AND THE CLIFFS' FURNACE

During the summer of 1867, a group of Eastern financiers organized a company at Norwich, Connecticut, to build a charcoal iron blast furnace, three miles northwest of Negaunee near Deer Lake. Edward C. Hungerford and John E. Ward were selected to supervise the construction and arrived in Marquette about August 1, 1867. Men were hired at once and land was cleared, and by late September a dozen houses were built and occupied, the foundation for the stack was completed and the wooden frame for the furnace was up, ready for the masons.

The site for this furnace (later called the Deer Lake) was on the Carp River at the foot of the Carp falls. It was a natural location for water power that, when harnessed by the company, turned an 18-inch turbine wheel with a 35-foot head, supplying all the power the furnace would need.[11]

About 50 men were employed at the site, quarrying stone for the stack, chopping wood (some coal pits were already burning), putting the road in shape to Negaunee, and erecting 12 more houses for the growing

[9] *Mining Journal*, October 24, 1874.

[10] *Ibid.*, February 20, 1875.

[11] The small diameter of the turbine wheel and of those later installed at this iron works and at Forestville suggests that Pelton type turbine wheels were used, instead of the water wheels like those which powered the earlier Carp, Collins and Forestville forges.

The Michigan Iron Industry museum. The Carp River falls is located aprox. 3/4 mi. upstream from the

He visited here in 1992 with E. C's daughters antoinette granddaughter as Bob
in R.V. trip   was E. C's son Roberts grandson
11   dau Bess m. but had no children

Deer Lake Iron Company, Ishpeming.
Note flumes for water power.

crew. The stack was up to 20 feet in late October and in March of 1868, the furnace was finished and would have been in blast except for a delay in receiving machinery.

On July 9, 1868, articles of association were filed for the Deer Lake Iron Company. The capital stock was set at $75,000, in 3,000 shares at $25 each. Investors were George P. Cummings of Marquette, Edward C. Hungerford of Chester, Connecticut, John E. Ward, James Lloyd Greene, James C. Colby, Daniel T. Gulliver, William R. Potter, Gardner Green, Caleb B. Rogers, Moses Pierce, Samuel B. Case, Theodore T. McCurdy, and Enoch F. Chapman of Norwich, Connecticut, and James H. Mainwaring of Montville, Connecticut.

The Deer Lake Furnace went into blast about September 1, 1868, and being the smallest stack in the Upper Peninsula — with a height of 33 feet and a bosh of seven feet, eight inches, its average daily production was only 11 tons. The furnace was operated six days a week, banked on Saturday at midnight until Monday morning, as the owners chose to observe the Sabbath. But by doing so, the furnace used more coal and made less iron and was not able to recover from the shutdown until the following Wednesday. For the year 1868 the furnace produced 600 tons of iron.

The furnace, though small, operated much more efficiently than most of the other stacks then in blast in the Peninsula. A cone-and-thimble affair at the stack top collected the hot gas which was piped to the hot blast oven and burned, heating it red hot. All of the gas was used for this purpose except that lost during charging. The hot blast oven was built on the Player's plan and contained castings weighing a total of 45 tons; it was large enough to satisfy a furnace of twice its size and was described as "enormous." Production for the first three weeks was sustained with charcoal made of one-half pine tops and the rest hardwood, with hard ore from the New York Mine which averaged 66 percent iron.

Earlier, when the company constructed a dam at the falls, it built a flume that ran down to the turbine. Water rushing down the flume, onto the plates of the turbine, turned the wheel with such force that 65 horsepower were developed through the main shaft. This, in turn, operated the crushers and hot blast blowing cylinders. In the 1870s, this same turbine provided the motive force for three circular saws that cut many millions of feet of lumber, from the choice "cork pine" logs, hauled from the company's holdings in the big Dead River valley pinery, lying north of Deer Lake.[12]

On September 3, 1872, the furnace wall broke out under the tuyere arch, and the molten metal ran out on to the floor of the engine house, causing a fire that quickly destroyed the casting house and the stack house. The timber skeleton around the outside of the stack caught fire, and the flames raced up the tinder dry wood, catching and burning the entire stack house. The furnace machinery was not badly damaged, and although the flume was destroyed, along with the mechanical connection between the water turbine and the blowing cylinders, a Mr. Hall, the superintendent, managed to have the furnace in blast again on October 19.

The amount of pig iron made and shipped from the Upper Peninsula blast furnaces increased yearly after the Civil War and in 10 years the combined total of iron, manufactured by the close of the 1867 season, was over 116,000 tons. This increased tonnage resulted in nearly doubling the cost of fuel by 1868. The average price for charcoal then was 11 cents a bushel, and the iron works were already finding a shortage of maple, from which most charcoal was made. Yellow birch was next in popularity, and a third of the furnaces were using pine, hemlock, and whatever softwoods were at hand. The amount of charcoal used in the stacks varied widely, and usually 110 bushels of the best grade would smelt a ton of iron, while that made from softwood required 140 bushels.

Just prior to June of 1867, the Iron Cliffs Company, owners of the Pioneer Furnace at Negaunee, had excavated and put in the foundations and retaining walls for their new Cliffs Furnace. Located about three miles southwest of Negaunee on what was then the popular Iron Cliffs Drive, it was situated on the west side of Foster Lake, about 50 feet from the shore. The ruins of this furnace are on the edge of the present-day Tilden Property of The Cleveland-Cliffs Iron Company.

Lumber for the furnace buildings and stone for the stack had been delivered, and work had been started on six of 14 kilns planned. The largest coal shed in the county stood ready for the first blast, but it would be many years before charcoal would be drawn from it. On June 10, 1867, the first train over the Mineral Branch of the Chicago & North Western Railway arrived at the location with fire brick, and eight masons started work on the stack.[13]

★ ★ ★ ★ ★

Because of financial troubles, the Northern Furnace at Harvey was shut

[12]*Mining Journal*, June 29, 1878.
[13]*Lake Superior Mining & Manufacturing News*, June 22, 1867.

down in May of 1867, and coaling operations were stopped in the woods. The furnace had produced 15,068 tons of pig iron in 207 weeks of running, and the outlook was that a long period of idleness lie ahead. In September of 1872, work was started on the Northern to refit it as a bituminous furnace, with the completion date set in two months time. But, in August of 1873, the work was still going on and dredging was underway of a 14-foot deep channel in the Chocolay River to receive ore and coal by boat and for shipping iron.

By January of 1874 the furnace work was nearly completed. The stack had risen to 50 feet, and the hearth and boshes had been enlarged to 52 inches and 13½ feet, respectively. The hot blast was also enlarged and changed to what was known as the Pennsylvania pattern, having 24 pipes measuring 4 inches by 12 inches, and 12 feet high. A water balance was made to hoist the ingredients, and three 31-foot-long boilers would supply steam for the hot blast engine.[14]

Though the furnace was about ready for the blast and the dock was loaded with bituminous coal, the decision was made to abandon the attempt to make iron, as the iron market was then in the clutches of the panic that started in 1873.[15]

JACKSON IRON COMPANY AND THE BAY FURNACE COMPANY

The Jackson Iron Company chose an ideal location on the Garden Peninsula in Delta County for its works, where 12,000 acres of choice hardwood lands were purchased, and where native limestone was found in abundance. The site offered many advantages — one of which was the natural sheltered harbor on Lake Michigan — which gave access to the lower lake ports. Construction on the No. 1 stack was started in February of 1867, and was completed in June of the same year, but delays in receiving necessary machinery postponed the first fires.

The furnace location on Snail Shell Harbor was first named Brownsdale after the company agent, Fayette S. Brown, but at his insistence it was later changed to Fayette.[16] In September of 1867 the company had almost 200 men at work and the total population of Fayette was estimated at from 400 to 500. Two streets, on which a dozen "neat" frame buildings were erected, circled the little bay, and on the opposite hillside stood a group of nearly 30 log houses. The location also had the familiar company store, a blacksmith shop, and some wagon shops.

By August of 1867 the company's tug, the *Rummage*, had towed several barge loads of iron ore from Escanaba, delivered there from their Jackson Mine in Negaunee over the C&NW Railroad.

[14]*Mining Journal*, January 17, 1874.
[15]*Ibid.*, May 3, 1890 (Note: Further evidence to support this is shown in the *Mining Journal*, January 29, 1876, which states: ". . . thousands of cords of wood . . . has been allowed to rot in great heaps . . ." in the woodlands surrounding Harvey.)
[16]*Lake Superior Mining & Manufacturing News*, September 28, 1867.

The first iron produced by the Jackson Furnace was run out on Christmas Day of 1867. With a hot blast pressure of two and one-quarter pounds, the furnace was soon making up to 27 tons a day, producing 155 tons during the first week of February, 1868. The founder at this time became aware of some damage to the lining of the stack, and lowered the hot blast pressure to one and one-half pounds, cutting production to about 20 tons a day, but to no avail as the furnace blew out for a new hearth on February 4. Making iron too fast was the reason given for the short 40-day blast, but the masonry work to reline the hearth was started immediately, and the furnace was soon in production again.

FAYETTE

The Jackson Iron Company started work on its No. 2 stack at Fayette in July of 1869 — a near duplicate of No. 1 with a nine and one-half foot bosh. The machinery for the stack was made at Cuyahoga Iron Works in Cleveland, and arrived aboard the steamer *Rocket* in September. The new buildings to house the machinery and the casting house were all made of stone and had corrugated iron roofs, while the stock house, containing a Blake Crusher for the ore and flux, was all wood. A shop over the stock house was furnished with a planer, lathe, drilling machine and other gear, and a feed mill that ground grain — all belt driven and powered by a 40 horsepower Root's steam engine.

MARQUETTE COUNTY HISTORICAL SOCIETY

Jackson Company's furnace, Fayette

Near the harbor stood a row of 11 top loading kilns with arched roofs, each measuring 14 feet by 18 feet and having a capacity of 75 cords of wood. Three miles northwest of the furnace were the Gates' kilns, a group of six round kilns, each 28 feet in diameter, 30 feet high, and able to hold 65 cords. Four other kilns were situated two miles southwest of the works, each having a capacity of 70 cords. Both of these kiln stations had log houses and barns for the woodsmen and animals. The furnace location also had a lime kiln that made 160 barrels of lime per week from the limestone out of the company's quarry.[17] There was also a 900-foot long dock and a saw mill with a daily capacity of 10,000 feet of lumber. Fifty houses, mostly log, were up by November of 1869, and a carpenter shop and a stone shop had been added.

About 250 men, many horses, and five yoke of oxen provided the labor at the iron works, where temperance was encouraged by the management and over-indulgence meant loss of a job.

On May 2, 1870, No. 2 stack at Fayette was blown in, but shortages of coal and repairs kept its production low for the first year. For the year ending November 30, 1871, both stacks produced a total of 8,696 tons of pig iron, an increase of over 2,000 tons from the previous year.

ONOTA

In the spring of 1869, a group of Marquette men organized a company to build a blast furnace six miles west of Munising on the shore of Lake Superior. As no land route existed for the transportation of heavy machinery, a dock was built into Lake Superior for receiving the furnace materials, and for the eventual shipping of iron. On July 9, 1869, articles were filed for the Bay Furnace Company, allowing a capital stock of $150,000, through the sale of 6,000 shares at $25 each. Incorporators were George Wagner, Jay C. Morse, Frank B. Spear and James Pickands of Marquette, William Shea of Munising, John P. Outwaite of Ishpeming and John Outwaite of Cleveland.

In August of 1869 the company began blasting at Powell Point for rock to be used in the construction of the 45-foot-high stack with a bosh of nine and one-half feet. The Merritt & Osborn Foundry of Marquette made the two 36-inch diameter horizontal blowers, and a concern in Cleveland supplied the driving engine, having a 16-inch bore and two and one-half foot stroke. Both arrived in September and were large enough for two stacks. The company owned 20,000 acres of choice hardwood lands, and for coaling 14 kilns were built at the location and two kiln stations with six kilns each were being put up in the woods two and one-half miles from the works.

The Bay Furnace location, first called Bismark then Wayne's Mill, was finally named Onota.[18] By December, the huge furnace, having a base 47 feet square and top 35 feet square, was complete. There were also a top house, saw mill, blacksmith shop, and some houses. The boilers and hot blast were placed on top of the stack to cut expenses in castings and pipes. Construction was started on the stack on July 19, 1869, and on March 5, 1870, the first cast was made.

The blast was started with 35,000 bushels of coal on hand, 8,000 cords of wood ready for the kilns, and pit charcoal being made in many places.

---

[17]*Mining Journal.* May 24, 1879.
[18]*Ibid.,* December 11, 1869.

In April, 17 log houses and eight two-story frame dwellings were occupied, a wagon shop was in business, and a school house was under construction.[19]

The furnace made about 20 tons of iron per day until May when it was shut down for lack of charcoal, after producing 800 tons of iron. To add to their troubles, 1,000 cords of maple were destroyed by a fire.

THE BANCROFT

IRON COMP'NY.

MANUFACTURERS OF

CHARCOAL PIG IRON,

FROM

Specular and Hematite Ores,

OF

MARQUETTE COUNTY.

PETER WHITE, President, Marquette, Michigan.
SAMUEL L. MATHER, Treasurer, Cleveland, Ohio.
JAY C. MORSE, Secretary, Marquette, Michigan.

Beard's Directory of Marquette County, 1873.

---

[19]*Mining Journal*, April 16, 1870.

Chapter V

THE TYPICAL U.P. BLAST FURNACE
IN OPERATION

The process of smelting iron ore — the reduction of the ore to a metal ready for the foundries and rolling mills — required large capital expenditures for labor and fuel. Labor was costly because the furnace required around-the-clock attention of a crew of men to keep the stack charged to its fullest at all times. Fuel was the greatest expense because large holdings of land were needed to provide the wood, as were choppers, teamsters, kiln burners, and expensive horses and mules, to make and deliver the thousands of bushels of charcoal used daily at each iron works.

As in the early forges in the Upper Peninsula, the ore was crushed, but there the similarity in making iron ends. The forge fires, never hot enough to melt the iron to a flowing temperature, merely fused the metal into a solid mass which had to be lifted out of the fires. The blast furnaces with their elaborate stacks, ovens, hot blasts, and many more distinct features, melted the iron which enabled it to be tapped and run off into the sand moulds in the floor of the casting house.

After being crushed, the ore was weighed and mixed with flux in measured amounts, depending on the quality of each ingredient. To each 800 to 1,000 pounds of ore, 50 or 60 pounds of flux would be added, and together they would be lifted to the top house by elevator (or "water balance") and thrown into the "tunnel head." At some furnaces the crushing, weighing and mixing was done in the top house and lifts were not needed. The ore and flux, with about 30 bushels of charcoal, made up a "charge," and a "tower" consisted of from 20 to 30 charges, from which two runs of iron were made in 12 hours. The charges were made in alternate layers; coal going in first, followed by the ore and flux, and each tower produced about six to seven tons of pig iron.

In the stack the different ingredients performed equally important functions in producing the final product. The charcoal provided the carbon necessary to displace the oxygen in the ore, and heated the metal and slag to a flowing temperature enabling it to run down to the hearth where it collected until tapped. The limestone flux united the impurities of the ore with the dross from the molten metal, and together formed a brittle glass, also a light glass froth. The founder or furnace hand in charge of the tower could tell the quality of the iron before it was tapped by observing the cinder being run out.

The hot gas generated by the burning charcoal was drawn off near the top of the stack and piped to heat the ovens of the hot blast. The gas also went to the boilers of steam engines which powered the blowing cylinders and various pumps around the iron works. At some iron works, like the Collins and Bancroft, water power operated the blowing cylinders or bellows.

The stacks in which the charges were placed were from 30 to 60 feet high, with boshes measuring from eight feet to 12 feet in diameter, but the early experimental Collins was much smaller in both respects.[1] The walls of the furnace were built of stone, except for iron-shelled stacks such as the rebuilt Marquette & Pacific, and measured at least 18 inches thick while the circular inside was lined with fire brick. The fireproof lining or hearth was carefully made to last as long as possible, and it was not unusual for a lining to last well over a year. But on many occasions, several months of production were enough to wear it out. The period during which the lining held up was called a "blast" and the furnace operated at "full blast" 24 hours a day until forced down by a worn lining or hearth.

The air from the blowing cylinders passed through a large pipe into the wind receiver or regulator, up to where it connected with the hot blast. It then went into the oven, where it was heated and passed down to the bottom of the stack, entering the furnace through the tuyeres to fan the burning charcoal.[2]

The molten metal was tapped from the furnace every few hours and run into the nearby casting house. Here, long channels in the sand floor called sows guided the red hot metal into the many smaller lateral moulds called pigs.[3] When the tap was complete the casting room floor was nearly covered with red hot shimmering lines of metal. After a short cooling time, men with sledge hammers broke the pigs from the sows, and wheeled them out to the stock piles and loading docks, hurrying, to be ready for the next cast.

The Lake Superior Iron Company, owners of one of the oldest, richest and most productive mines on the Marquette Range — the Lake Superior at Ishpeming — began construction on an experimental blast furnace in the summer of 1871. It would test the smelting ability of peat in use with the leaner ores found in great abundance in the area. Corporators of the company were Heman B. Ely and Anson Gorton of Marquette, and Samuel P. Ely, George H. Ely and Alvah Strong of Rochester, New York.

Articles of association were filed on March 13, 1853, which allowed $300,000 in capital stock. From then to 1871, mining was the principle aim. Land holdings included 200 acres of a large peat bed lying in Ishpeming, over nine feet in depth, which yielded 3,000 tons per acre.

[1]Gay's experimental furnace stood only eight feet high, had a bosh of two and one-half feet, a hearth 15 inches square, and tunnel head 12 x 15 inches, where it was filled — according to the *Lake Superior Journal* of February 6, 1858.

[2]*Mining Journal*, August 27, 1870.

[3]Fisher, Douglas Alan, *The Epic of Steel* (New York, 1963), p. 31.

Blast furnaces in Eastern states had used dried peat as a fuel for smelting iron ore with good results for some time prior to the discovery of the peat bed in Ishpeming. As a result, the Lake Superior Iron Company initiated the Lake Superior Peat Works in 1870, and on July 1 of that year started processing the raw peat for use in a furnace which was soon to be constructed. In and about Ishpeming dried peat was used in homes for fuel in heating and cooking and served the purpose well.

On startup, the peat works used racks exclusively in the drying process. A mixing mill was set up that filled four of these racks per minute and in the course of a day 1,000 racks were filled and set out to dry in the sun. After five or six days in the open the racks were carted to sheds where they dried for an additional seven to twelve days. Twelve men handled the racks and each day 1,000 were stored in the sheds for drying. The storage sheds had room for 15,000 racks but how long the peat works operated before the company's blast furnace was built is unknown.[4]

Original plans were for a stack 22 feet in height with a bosh of four and one-half feet, but during construction the company decided to make it almost as large as the Deer Lake furnace. In March of 1872 the furnace was completed and put into blast using charcoal and ore from the Lake Superior Mine. One and one-half tons of metal were made in the first cast, and as the stack gradually warmed up the charcoal was replaced with dried peat. But it was soon found that peat alone did not produce the desired results and the blast was stopped. A number of kilns were then built to provide the necessary amount of charcoal needed with the peat, and in July it was put in blast using 25 percent charcoal and 75 percent peat — but production was still very low.

The peat furnace was located south of Ishpeming near the Marquette & Ontonagon Railway and adjacent to the peat works, where in August of 1872, 40 men were employed digging, tramming and preparing the peat for the furnace.[5] The soggy black peat was first shoveled from the beds, placed into small cars that held 750 pounds each and pushed on an inclined tramway up to the grinding mill. There they were emptied into a hopper where the wet mass was ground up into a uniform size, then loaded into tram cars again and pushed to one of the many drying sheds where it was spread out in a layer about five inches thick. While drying, the peat cracked into small pieces, like a giant puzzle, just the right size for charging in the furnace. The drying process was completed in seven to 10 days. The company had a saw mill set up at the iron works and the lofty pines cleared from the peat bed provided more than enough lumber for all of the necessary drying sheds, which together housed over 70,000 square feet of drying space.[6]

[4]*Negaunee* (Mich.) *Mining Review*, July 28, 1870.
[5]Located on the side hill near the present Pine Street and Excelsior Street intersection.
[6]*Mining Journal*, August 3, 1872.

East Munising
Furnace in
blast, 1870.

The small stack, having a bosh of only 6½ feet, was charged with only peat in October, but the resultant cut in production — down to about eight tons per day — prompted the company to blow it out for enlargement. The stack height was increased to 48 feet, the boshes widened to nine feet, bigger boilers and hot blast added and, on October 1, 1873, it was charged with charcoal and put into blast. Not long after, peat was again introduced in a ratio of 120 pounds to 400 pounds of charcoal and, with the hot blast pressure at three quarters of a pound, it was making up to 10 tons per day and increasing daily. The peat furnace continued to make iron until the financial panic affected its iron sales and in 1874, after producing 1,150 tons of iron, the furnace was shut down.

THE FIRST IRON STACK ON THE PENINSULA

In the summer of 1871, three locations on Dead River — Forestville, Stone's Mill (where one upright and five circular saws were cutting a total of 20,000 feet of lumber per day) and probably Collinsville — provided $1,200 to build a public school at the central location of Forestville, high on a hill behind the iron works. The school, with a large masonry basement, was built by the people of Forestville, under the direction of S. R. Gay. It was supported financially by the three locations, and was independent of the school district.[7]

The 40-foot stack of the Bancroft Furnace at Forestville was then undergoing changes that for a while made it the most modern furnace in the Peninsula. Twenty feet of the Bancroft's stone stack was taken down and replaced by an iron shell, setting the pattern for some later stacks, and in December of 1871 the furnace was again working, making 110 tons per week.

[7]Seventy-five pupils attended this school; the foundation is still visible.

Bancroft Furnace, Forestville, c. 1872. Hotblast pipe runs from building on Dead River where ovens and water power plant were located. The village had a schoolhouse (top center) with three smoking kilns below.

Charcoal for the furnace was supplied from 16 kilns, with a total output of some 12,000 bushels per week. The kilns were scattered in the hardwoods about the hamlet, with three located next to a bluff just below the school house. In the early 1870s nearly 100 men and 24 teams were employed by the concern in the iron making and saw mill operations.

The *Mining Journal* carried this account of a foreman's narrow escape at the furnace in March of 1874:

> The furnace was acting like an animal does after taking a large dose of caster oil and was pretty soft inside, but rather hard at the forebed; and as he was trying to ease her of her burden, she flew at him like a fiend. He succeeded in getting a hole in her and after pulling the bar out, the cinders flew like water from a hose, striking him on the shoulders, back and legs, burning his pantaloons badly. But as he was quick in getting them off, he escaped with little or no injury.

Forestville, unlike other furnace locations where John Barleycorn was vigorously opposed, boasted a newly built saloon in the winter of 1874 (at least the men were gratified at its existence) called the Silver Lead, owned and tended by a German.[8] A harness shop and a shoe shop were also erected about the same time. The boys at the iron works took advantage of this oasis in the hardwoods and in a short time their spirited behavior was the object of scorn by the upright citizens of Forestville.

Temperance societies were then in vogue in Marquette and surrounding communities, and they bore down on those not practicing sobriety. A temperance society was formed in the little hamlet to fight the evils of the Silver Lead.[9] The temperance union meant business, even to the extent of marching through the streets of Marquette. But to their dismay,

---

[8]*Mining Journal*, February 28, 1874.
[9]*Ibid.*, March 21, 1874.

the group suffered an irreparable break in its ranks when the president of the society moved to Marquette to run a saloon![10]

The Bancroft produced 3,556 tons of iron in 1874 and continued to make iron until May of 1876. That month the Dead River overflowed its banks and flooded the riverside iron works, destroying the flume and causing the furnace to be banked up. An issue of the *Mining Journal* in February of 1877 states, "... there is but one furnace in this county blowing, the Marquette & Pacific ...", and by 1880, the Bancroft furnace at Forestville was dismanted.[11]

The Lake Superior Iron Company, owners of the peat furnace in Ishpeming, started construction on another furnace in January of 1872 near Marquette harbor on the site now occupied by the Marquette Dock Company at the foot of East Washington Street. The furnace was designed to burn anthracite coal, the first such in Michigan, and was completed in December of that year. The blast engine, built by Robinson, Rea & Company of Pittsburgh, had a steam cylinder of 34-inch diameter, which powered an 84-inch diameter blowing cylinder, both having a 4½-foot stroke with a total weight of 70 tons. It was made to operate at 25 revolutions per minute, turning two 10-ton flywheels.

The stack was the same height as that of the Marquette & Pacific, 60 feet, but had boshes of 17 feet and used an iron shell, built in Pittsburgh, which was supported on eight cast iron columns. In late November of 1872, charging was started with 40 cords of wood put into the stack, then 50 tons of coke were added. This was followed by regular charges of ore, fuel and flux and, on the evening of December 10, a group of gentlemen gathered around the furnace to watch Samuel P. Ely light the fire. At 9:30 p.m., all was ready, and telling the group that he named the furnace Grace, after his youngest daughter, Mr. Ely applied the match. The December

MARQUETTE COUNTY HISTORICAL SOCIETY

Grace Furnace, Marquette, ready for casting, about 1872.
Stack was constructed of iron.

[10]*Mining Journal*, May 9, 1874.
[11]*Iron Agitator*, Ishpeming, June 12, 1880.

14, 1872, *Mining Journal* went on to explain, ". . . The party then adjourned, at Mr. Ely's suggestion, to the engine room, where upon the elevated floor around the ponderous blower, they partook of a bounteous supply of oysters and coffee, and afterwards, amid the wreaths of cigar smoke spent some time in social chat . . . ."

The fire burned on its own until the next afternoon and the blast was started. The following week the Grace Furnace was making 30 tons of metal per day. In May of 1873, defects in the design of the boshes and poor brick work caused the furnace to be blown out. Masons were then hired to make the repairs and in July she was put in blast, going on to make from 40 tons to a high of 63 tons per day. The Grace continued in this blast until March 24, 1874, when it was blown out after making 9,376 tons of pig iron. This marked the end of the Grace Furnace and, although it was later leased by two different concerns with intentions of putting it back in blast, production was never resumed.[12]

When the anthracite-burning Grace began making iron in 1872, a considerable amount of thought was given to building yet more furnaces in Marquette, along with mills, foundries, steel works, rail mills, locomotive works, and much more. A member of Samuel Ely's party at the lighting of the Grace said, ". . . The iron interests of the Lake Superior is yet in its infancy . . . . We have this furnace and we will build more . . . let us resolve to build more furnaces . . . ."

Peter White showed more interest, probably, than anyone in the future development of the iron industry in Marquette County at that time, but even though he was heavily invested in local iron companies, there seemed to be no concerted effort to make Marquette another Pittsburgh. Peter White introduced legislation in the Michigan State Senate in 1875, allowing incorporation capitalization to increase from $500,000 to $4,000,000. At the time the country was in the throes of a depression that had started with the panic of 1873 and continued to affect the iron industry for many years thereafter. By the time business picked up again coke iron was in demand. Other important reasons why iron-centered industrial development did not occur in Marquette were the continued expansion and improvement of the Bessemer steel making process and the fact that Marquette was not in a good position for marketing the metal.

THE ESCANABA FURNACE

The Cascade Iron Company, owned by William Bagley, Joseph Kirkpatrick, William Smith, Samuel Hartman, Samuel Riddle, and James

---

[12]The *Iron Agitator*, July 31, 1880, reported, "Hon. John Burt, has taken a lease of the Grace furnace for himself and his associates. . . ." The *Mining Journal* (Weekly Edition), September 16, 1882, noted ". . . The Grace furnace has been sold to Sturges, Kruse & Travers who will make an effort to secure the fuel and make repairs necessary to put the stack in blast the coming winter. . . ."

Lyon, purchased 3,120 acres of mineral-rich land near Palmer in 1869, then organized the Escanaba Iron Company for the purpose of making pig iron. A site was chosen one and one-half miles north of Escanaba on Bay de Noc and in the fall of 1871 construction was started.

Variously called the Cascade and the Escanaba, this was supposedly the "largest charcoal furnace in the United States" at the time. Near the close of 1872 or early in 1873, the furnace was put into blast and, like other stacks of the Peninsula, it was destined for a short life — even though it was of the most modern design and had the best machinery. The cheap, lean ores abounding on the Marquette Range were attractive to the company and, in a reckless attempt to save money, these ores and a poor grade of charcoal were first used which caused a scaffold chill and forced the furnace to shut down. When it was placed in blast again the ore and fuel were of the best grades available.

The 56-foot-high stack with 12-foot boshes was mounted on columns, and three tuyeres directed the hot blast into the 48-inch hearth. While the Grace in Marquette was ringed with seven tuyeres, most on the Peninsula were fitted with two or three.

In May of 1874, the Escanaba was making from 30 to 35 tons per day and, by October, 49 kilns were supplying the charcoal, but not in sufficient quantity to sustain its full iron making ability. The kilns were situated at Kloman, Perkins, Maple Ridge, and Mile Post 22, all along the C&NW Railway. Ten kilns of 50-cord capacity were at Kloman, 13 at Perkins where five held 30 cords, and the others had a 50-cord capacity. At these kilns, 125 men were employed chopping, hauling and burning the kilns. At Mile Post 22, 14 kilns of 30 cords each kept 30 men busy, and at Maple Ridge there were 12 white bee-hives of 30 cords employing 100 men. In all, 330 men worked at the kiln stations and 50 more were employed at the furnace. George English, company agent, was then trying desperately to recruit 300 more choppers to harvest the 40,000 cords needed for the coming year.[13]

The production of the Collins Furnace through the years averaged from nine tons to a little more than 11 tons per day. Output was affected by the physical condition of the works which was the responsibility of the appointed general manager. During this time, the company subjected the furnace to the whims of different iron makers — some were good, others of dubious skill — and for years the furnace suffered for it. While one was conscientious, kept machinery in repair, stack properly lined and the kilns burning steadily, a replacement would concentrate on production, and gradually run the iron works into the ground. This went on for some time until a Mr. Larned was installed as the superintendent. Larned ran a tight ship and the Collins then went on to make 4,631 tons of pig iron in 1867, and in a six-year period it did not have a blast lasting less than 10 months.

[13]*Mining Journal*, November 21, 1874.

This remarkable record ended in May of 1873, when it blew out for the last time after the company lands had been stripped of all charable wood.[14] It was said that almost $2 million worth of pig iron were made in this water powered furnace on Dead River.

The machinery was all removed from the old Collinsville Furnace and in 1889, Mayor Clark of Marquette headed an effort to acquire the valuable land the furnace occupied to erect the city's first hydro-electric plant. Clark, in negotiations with R. K. Hawley of Cleveland (then owner of a track of land along the Dead River containing an estimated 200,000,000 feet of pine timber), agreed on a sale of 400 acres at Collinsville including all the old wooden frame buildings, from which the hydro-electric station was built later that year.

During the week of August 18, 1889, the razing of the furnace and buildings was started. Hawley retained all the fire brick and stack lining from the blast furnace and hauled it to the mouth of the river. Here it was used in the construction of the boiler fireplace and stack for the huge steam sawmill then being built at the present site of the Upper Peninsula Generating Company.[15]

PENINSULAR IRON COMPANY

About three miles south of the Collins location, at the mouth of the Carp River, John Burt began construction of a blast furnace in the summer of 1872. It had been his dream for many years to locate a furnace there. In 1851, he had purchased 2,600 acres of valuable iron and timber lands, most of it lying three miles west of the Jackson Mine in Negaunee, and 500 acres with one-half mile frontage on Lake Superior, which controlled the water power of the Carp River. Burt and 15 men then dammed the river, and put up a water-powered sawmill which was operated for many years. An elaborate iron works was to be built next but the plans for this were long dormant.

In September of 1872, it was reported that Burt was more determined than ever to erect a furnace on his lake shore property, and in October work on the 10-foot bosh, stone stack was begun. Before the furnace was completed stockholders of the company met in June of 1873 and elected John Burt president, Hiram A. Burt vice president, and Noah W. Gray secretary-treasurer. The company was authorized to sell $250,000 in stock. The Carp River Iron Company was consolidated into the Peninsular Iron Company in January of 1874. On April 24 of that year the stack was filled and the furnace cast its first iron two days later. Charcoal for this run was made in pits on Mesnard Hill and in a number of kilns near the furnace.

Shortly after starting, the hearth in the furnace proved defective and

---

[14]*Mining Journal*, May 31, 1873.
[15]*Ibid.*, (Weekly Edition), August 10, 1889.

10 tons of iron ran out on the ground. Three months later, in September, it was back in blast but blew out in December because of the poor iron market.

The Peninsular Iron Company operated the furnace that first season then leased it to Hiram Burt in 1875, who ran it sporadically until 1882. In 1878, a bell and hopper was added and that summer, 16 kilns at the "Dutch (German) settlement" south of Harvey, with the kilns at the location and those leased at Champion, supplied the charcoal. The August 2, 1879, issue of the *Mining Journal* relates that "Louis Koepp is hauling charcoal day and night (from the Dutch settlement), and is always as black as a brunet African."

MENOMINEE FURNACE

J. W. White of Clarksburg took on the responsibility of superintending the construction of a blast furnace at Menominee ". . . built under the auspices of the National Iron Company of Depere . . . ,"[16] on which work was well under way in September of 1872. Located a short distance north of the town and called the Menominee, this furnace was intended to run on pine slabs and edgings, a waste by-product of the many saw mills strung out along the mouth of the Menominee River, and to be charred in a number of kilns situated about the furnace. The use of wastes to make coal was not new in this district, as the Michigan and the Champion furnaces were run partly on their sawmill scraps for some years.

The 44-foot-high iron stack with boshes of 9½ feet went into blast about July 8, 1873, and on the fourth day running it was making 10 tons per day. The hot blast steam engine, with a piston of 18-inch diameter and 28-inch stroke, ran at 57 RPM's with a head of 65 PSI of steam, and operated a blowing cylinder of 48-inch diameter with a five foot stroke. Richard Dundon, ". . . one of the best founders of the northwest, and one of the brothers of that name, all of whom [had] enviable reputations as iron makers . . . ," was the founder, and started up with 175,000 bushels of charcoal on hand.[17] To be sure of a steady supply, Dundon contracted with the farmers along the Chicago & North Western Railway to make pit charcoal.

The Menominee suffered the usual fluctuations during the panic, running out of an iron market, running out of ore, running out of fuel, etc. In 1875, the company built a kiln station 22 miles north at Stepenson to make hardwood charcoal as the mill wastes were not sufficient. During the summer of 1876, the furnace was completely overhauled with a new hot blast oven and a bell and hopper added to the stack, and went into blast with 220,000 bushels of charcoal stocked.

The *Mining Journal* related that, for the week ending April 21, 1877,

---

[16]*Mining Journal*, August 17, 1872.
[17]*Ibid.*, May 31, 1873.

the furnace ". . . made 301½ tons weighing 2,268 lbs. to the ton . . . and made 2,397 pounds of furnace castings . . ." using:

Cleveland granular ores	875,885	pounds
Winthrop hematite ore	216,715	"
Slacked lime (flux)	55,800	"
Charcoal	34,925	bushels
Kinds of coal — maple	16,648	"
soft	4,665	"
mixed	3,917	"
Pine, hemlock and black ash coal from home kilns	3,696	"
Pounds of blast pressure	4	PSI.
Temperature of oven	829	degrees F.

The Menominee Furnace came on hard times later in 1877, banking up in September and finally putting the fires out in late October to remain idle all that winter. In May of 1878, a Mr. Cherry and a Mr. Cox, agents for the Leigh Coal Company, leased the iron works with the kilns at Brookside, Kloman and Stephenson, and renamed the furnace "Champion."[18] In July, the Champion started making iron.

On December 4, 1880, the furnace was sold in a bankruptcy sale to A. B. Meeker of Chicago, who rechristened the stack "Menominee" in January of 1881. It was then making up to 30 tons per day. On December 11, 1882, it was blown out for repairs after a run lasting more than a year, during which time 11,377 tons were cast, including 10,400 tons in 1882. It went into blast again in January of 1883, but immediately there was talk of shutting down because of poor iron prices. Ores from Florence, Norway and the Great Western Mine were then being used and in September of that year it was still in blast, making 31 tons per day.

The Menominee Furnace was sold at auction to Eric L. Hedstrom for $17,000 in May, 1889, and it was then stated that the furnace was outdated and could not compete in the industry.[19]

FURTHER DEVELOPMENTS AT FAYETTE, DEER LAKE

The original railroad from the Fayette Furnace to the Jackson Company's kilns was laid with wooden tracks. But by 1872, this 5.75 mile railroad was relaid with 28-pound iron rails. It had two steam locomotives and 23 cars for moving the wood and coal, and one of these engines was capable of pulling 16 loaded cars over the grades.[20] The railroad was later improved upon and extended to Sac Bay, some seven miles from the furnace.

An explosion in the No. 2 stack late in the summer of 1875 caused it to fall in, which stopped iron production for weeks, and on December 15 of that year a $50,000 fire occurred when molten iron broke over the furnace dam, stopping the flow for two months. Another fire, in

[18]*Mining Journal*, June 22, 1878.
[19]*Ibid.*, (Weekly Edition), May 11, 1889.
[20]*Ibid.*, May 24, 1879.

February of 1879, destroyed the Catholic church and a valuable collection of books that belonged to the pastor.

In March of 1880, a number of Canadian woodcutters were imported to keep the kilns full and, with these added to the payroll, more than 300 laborers collected nearly $6,000 each month in wages.

While drinking was discouraged at the Fayette location, within walking distance on the nearby wagon road and off the lands of the iron company, were saloons that catered to the desires of the isolated furnace hands and woods workers. Two of the closer saloons were "Pig Iron Fred's" and Jim Summers' "house of ill fame," the latter run by the hard-drinking leader of a bunch of toughs. In August of 1880, a young girl being held against her will at Summers' escaped and spent two days in the woods, then went to the furnace location to the deputy sheriff and asked for protection. The deputy, being a very timid man, made the girl return to the "house of shame," explaining to the furnace people he did not want Jim Summers mad at him. The furnace people had previous difficulties with Summers and this latest incident set people in motion for a "house cleaning." A public meeting was called and nearly 100 men gathered to rescue the forlorn girl, and to break up Summers' gang.

At the meeting someone reported that one of Summers' men was breaking things up and fighting at Pig Iron Fred's, and the furnace people moved. They walked the short distance to the saloon and as they came near, one of Summers' men was saying he could "whip any S.O.B. on this side of the bay." The men soon had him knocked to the floor where he was clubbed, kicked, stoned, jumped on and nearly killed by the enraged furnace hands. He was then taken to the jail in a butcher's wagon, ordered onto the first boat going across the bay, and warned that if he was ever seen there again they would hang him. The crowd of men then went to Summers' house. They whipped Summers and knocked him out, leaving him lying senseless in the woods.

The girls were removed from the building, and the men smashed kegs of beer, emptied bottles of whiskey, wrecked the furniture and they torched the building and watched it burn to the ground. Meanwhile, Summers revived and took refuge in the woods. Although 50 men spent two days looking for him, he was never seen again.[21]

★ ★ ★ ★ ★

The owners of the Deer Lake Furnace were confident demand would remain for their iron and in March of 1873 they started construction on a second stack. Unlike the stone structure of No. 1, this stack was an iron shell, being 45 feet high with a nine-foot bosh and closed top. The new stack went into blast in January of 1874 and in April it was making 20 tons per day.

[21]*Mining Journal*, August 7, 1880

That year, the Deer Lake Company dismantled the top half of the No. 1 stack and replaced it with an iron shell and a bell and hopper. Using slab-charcoal, it went on to produce 17 tons per day. At this time a tram road was in use from an Ishpeming rail connection to Deer Lake, one and one-half miles long, on which pig iron and ore for the furnace were transported by wagon. But, by 1878 (the furnace having been cold for some two years because of the iron market), the road was in a state of serious disrepair. Mules were no longer seen pulling the heavily-loaded wagons and cows used the kilns for shelter. The furnace buildings were stacked high with lumber from the sawmill.[22]

The Bessemer Branch of the C&NW was extended to the furnace in the early 1880s, and the *Iron Agitator* stated in September of 1884 that "The road to Deer Lake has done away with the hauling of ore and pig iron over the former highway . . . . The sad-eared Deer Lake mule that was wont to haul 'de charcoal from de kil,' lulled into dreamy insensibility of all his surroundings by the high keyed notes of the Canuck driver . . . ."

Water power operated the blowing cylinders of both stacks for some years. The water was fed to the turbine in a flume of three-inch plank, running high on a trestle straight from the dam. In 1880, two turbines were in place, one of 20-inch diameter operated the hoist and blowers and the other of nine-inch diameter ran the Blakes crusher. To assist these turbine wheels in the summer months when the river's flow diminished, two 40-horsepower steam engines were fired, and business went on a usual with the water power running only the sawmill.

Thirty-nine kilns were kept burning with both stacks working and 300 men were employed as teamsters, choppers, swampers, furnace-men, carpenters and colliers, producing iron and lumber. The location had 50 buildings, including barns and other out-buildings, and the houses were well maintained. In August of 1880, the Deer Lake Furnace and location passed into the hands of a corporation named the Deer Lake Company.

On the 28th of June, 1884, the first shipment of pig iron to be made over the the Marquette & Western Railroad (the grade is still visible on the hillside north of Mount Marquette skiing complex) reached Marquette from the Deer Lake furnace, and was shipped on the propeller *St. Paul* to Cleveland. The total product for the Deer Lake in 1884 was 10,753 tons, which was the best year ever to that time.

---

[22]*Mining Journal*, June 29, 1878.

DEER LAKE

Iron and Lumber Co.

MANUFACTURERS OF

CHARCOAL PIG IRON AND BLOOMS,

FROM PURE LAKE SUPERIOR ORES,

AT THEIR

Furnace near Ishpeming, Mich.

GARDNER GREEN, President, Norwich, Conn.

D. R. SULLIVAN, Vice   "   "   "

THEO. F. McCURDY, Treas'r,   "   "

C. H. HALL, Agent,   Ishpeming, Mich.

From *Beard's Directory of Marquette County*, 1873.

Chapter VI

THE END OF A CHAMPION

The Champion Furnace at Champion reported a total product of 5,560 tons of iron for the year 1870 when it was in operation only 10 months — a record comparable to any Lake Superior furnace of its size. An average of 70 pounds of flux were used per ton of iron, and the ore used was about half Lake Superior hematite and half Champion oxydes, red and black. The ore yielded 65.71 percent iron, with an average of 103 bushels of charcoal to the ton. Twenty pounds of charcoal was considered a bushel at the iron works.

The Champion Furnace was blown out in October of 1873 for extensive changes and rebuilding. The stack height was increased to 46 feet and the bosh to nine feet, with a closed top added for economy. Eight kilns were added, with a new hot blast and a Crane Brothers hoist. A long trestle was built to the new stock house, which measured 36 x 55 feet, and one of the popular Blakes crushers was installed. The furnace was back in blast in January, 1874.

Champion Furnace, long out of blast, From *Lake Superior Mining Institute* proceedings, 1903.

On Thursday morning, April 9, 1874, at quarter past eight, the Champion Iron Works suffered a devastating fire which did $25,000 damage to machinery and buildings. A tuyere burst and molten iron ran out on the floor, spreading fire throughout the buildings in a very short time, completely destroying the hoist, stock house and the engine house. Surprisingly, the stack, hot blast engine, and boilers survived the fire, and the stack was later shoveled out and considered fit for rebuilding.[1]

Although the stack could be used again and a large stock of cord wood and charcoal were saved, the Morgan Iron Company deciced not to rebuild because of a lack of insurance on the works and a faltering iron market. The furnace location was nearly deserted in July, 1874, and to salvage a part of its investment, the company auctioned off all the remaining equipment.

ONOTA

A second stack was planned in April of 1872 for Bay Furnace at Onota near Munising, and the following month the tug *Dudley* brought men and supplies from Marquette to begin construction on the iron shell stack. Billy Boals had the contract to put up 25 more kilns in the hardwoods, making a total of 52, and the Iron Bay Foundry of Marquette manufactured the blowing cylinders, shell, steam engines and other required machinery.

The cornerstone for No. 2 furnace was laid in mid-July, and on December 15, 1872, it was blown in, casting 23 tons per day within a few weeks. The economy of Onota was still very sluggish, even with both stacks in blast. But they were in good spirits and had dances during the winter months to entertain the women, while whiskey kept many of the men happy. With both stacks running, nearly 3,500 cords of wood had to be charred each month, which employed up to 150 men in cutting, hauling and burning operations. Ore was supplied from the Republic, Kloman, Cleveland, and McComber Mines in Marquette County, and was brought in by ship during 1874. Through the years of its existence, the Bay Furnace received its iron ore supply from the port of Marquette on the steamers *Ira Chaffee* and *St. Clair*.

Bay Furnace No. 2 stack was 45 feet high with boshes of 9½ feet, and at times cast as much as 35 tons per day. It started a blast on May 25, 1875, lasting 316 days, during which time it consumed 15,847 tons of ore and 1,045,440 bushels of charcoal — mostly hardwood — to make 9,695 tons of pig iron.[2] Charges consisted of 1,100 pounds of ore and 30 bushels of charcoal, plus flux. One hundred twenty of these charges were made in a 20-hour period. For the month ending June 18, 1876, it produced 1,000 tons of Bessemer iron.

[1] *Mining Journal*, April 11, 1874.
[2] *Ibid.*, May 6, 1876.

Since its inception, the well-planned village of Onota grew with streets on which 80 cottages were situated, and over 500 people lived, all dependent on the furnace. Henry S. Pickands, in charge of the furnace property, insisted upon order and a neat appearance, and the streets were cleaned and the dwellings whitewashed regularly.

The Catholic and Protestant population were well cared for. A Methodist minister resided at the location and there was a Catholic church erected at a cost of $3,500. Dr. H. D. Pickman watched over the health of the inhabitants, serving as doctor and apothecary.[3]

The Bay Furnace produced iron regularly until May 31, 1877, when Onota was entirely consumed by fire. It seems a teamster was bringing in a fresh load of charcoal hot from the kilns and, on his way through the village streets to the stock house, the wind fanned the coal into a red hot mass. Desperate to save the horses, he unhitched them from the burning wagon, leaving the flames to quickly spread to nearby buildings. The furnace, office, store, and houses were destroyed along with 60,000 bushels of charcoal. The only buildings to survive were the schoolhouse and the church. A Frenchman and his large family lived in the Catholic Church until October of 1879 when that, too, burned down, leaving only the schoolhouse standing.

Everyone managed to leave the burning village safely and eventually made it to Munising. For a while nearly 125 people found refuge in the Munising schoolhouse, where one of the frightened women gave birth to a child. The tug *J. C. Morse* was at the dock when the fire started and saved many people by carrying them out of range of the flames. The tug was credited with saving the lives of most of the people.[4]

There was the usual talk of rebuilding, but the village was not rebuilt at the original location. Several years later Onota was re-established as a charcoal producing center at a point about 20 miles west on the Detroit, Mackinac & Marquette Railroad. While the present ruins of the Bay Furnace consist of only one stone stack, the No. 2 stack was an iron shelled structure and was probably reclaimed as scrap iron.

THE CLIFFS IN BLAST

The *Lake Superior Mining and Manufacturing News* of Negaunee reported in December, 1867, that "The Cliffs is nearly completed. Work has been suspended for the winter." The company had spent $40,000 on the Cliffs that year and suspension of work was a long one, lasting until late in 1873. All new machinery was then installed, including boilers and a Mackintosh & Hamphill blower. The original blowers had been

---

[3]*Mining Journal*, January 17, 1874.
[4]*Ibid.*, June 2, 1877.

scavanged a few years earlier for the Pioneer, whose boilers had collapsed from overheating while in blast.

The Cliffs furnace was finally blown in on March 14, 1874, and two weeks later it was making 16 tons per day, with a total production of 3,540 tons for 1874. In and out of blast for about three years, the Cliffs produced slightly over 8,000 tons of pig iron before being abandoned. In March of 1879, the Pioneer Furnace was in trouble again and in order to put one of its idle stacks back in blast, it was necessary to remove and use all the machinery from the Cliffs.

THE ESCANABA DISMANTLED

An accident occurred at the Escanaba Furnace in September of 1874 that forced it down for some days. A furnace hand was charging the stack and while dumping his buggy he hit the heavy bell and hopper which was already loose in its settings, and it fell down into the stack. The hot blast was shut off immediately and the cumbersome extra charge was hoisted out, undamaged.[5]

The financially troubled iron company optimistically advertised for 100 men to work at Mile Post 22 in November, but the iron market declined and the furnace was blown out for the final time shortly after. The Escanaba Iron Company was adjudged bankrupt in March, 1875, and offered to settle its bills at 60 cents on the dollar. A. B. Meeker purchased the iron works at public auction in 1876, but in July the *Mining Journal* reported ". . . the Iron Cliffs company has purchased all or part of the wood belonging to the Escanaba furnace, and will have it made into coal for their own use . . . no prospect of the Escanaba stack being blown in."

All of the mechanical parts of the Escanaba Furnace were dismantled and moved to Pittsburgh in 1878.[6] After the equipment's removal, the iron market improved and the Escanaba's people bemoaned the loss of their furnace.[7] A new one was talked of but Stevens and Atkinson of Negaunee purchased the houses and brick walls on the location in April, 1879. Most of the houses were moved into town, and the brick was sold to builders. By October, when only a few houses remained, it was said again that it had been a grave mistake to tear it down.[8]

DONKERSLEY'S LEASE OF THE MORGAN FURNACE

Because of a dwindling supply of nearby hardwood, the Morgan Iron Company decided to blow out the furnace on November 17, 1868. It then concentrated efforts on making a wooden tramway to reach newly-purchased hardwood stands in the vicinity of the Dead River. The

[5]*Mining Journal*, September 26, 1874

[6]*Iron Agitator*, June 12, 1880.

[7]*Mining Journal*, November 30, 1878.

[8]*Ibid.*, October 4, 1879.

tramway was built at a cost of nearly $5,000 per mile and eventually extended north from the Morgan location, a distance of nine miles. Charcoal kilns were built at three different places along the road, the first set being six miles out, the second set two miles farther and the last set near the end of the road. The tramway passed through some extremely hilly country and considerable trestlework was needed across deep ravines. The steam saw mill of Decker & Steele located at Eagle Mills provided the hardwood rails used on the tramway. These were cut on the one circular saw. When completed, mule-drawn coal cars containing 500 to 600 bushels of charcoal, making two trips a day, were used to transport the fuel from the Dead River kilns to the Morgan Furnace.[9]

While the furnace was shut down a new hot blast was installed on one side of the stack, the casting house was enlarged, and a device was fabricated that used the waste heat coming out of the tunnel head to dry the crushed ore before charging the furnace.

The furnace was again making pig iron on December 28, 1869, after being out of blast for over a year. The new hot blast with other improvements proved to be very beneficial, as the production of one long ton of iron consumed only 98.66 bushels of charcoal of which one-fifth were made from softwood. The average weekly production was 134 tons.

An enormous salamander developed on the Morgan hearth in June, 1871, after the stock in the furnace dropped down and blocked both tuyeres. With the loss of the hot air blast, the mass soon chilled and spread across the hearth from one tuyere to the other to a depth of 45 inches. The usual method of removing these hardened masses was by pick and shovel and in some extreme cases blasting was even tried, but the inventive Cornelius Donkersley decided to try something never attempted before in an Upper Peninsula furnace. He proceeded to have a hole punched through the thick furnace wall above the salamander and installed a large tuyere. Coal oil was then forced under pressure into the tuyere from a pipe running from the top house and the blow torch effect soon had iron and cinder running out of the notches. Six days and seven barrels of oil later, the salamander had been entirely removed and the furnace was again running smoothly, making charcoal iron.[10]

Donkersley, with other investors, leased the faltering Morgan Iron Company in October of 1874, believing they ''. . . could make the production of pig iron pay in spite of the depression . . . that was then upon the land.''[11] This lease also covered the Champion coal kilns which were then being used to make coal for the Morgan Furnace.

In April of 1875, the colliers at these kilns, deprived of a store to trade

[9]*Mining Journal*, August 17, 1871.

[10]*Ibid.*, June 10 and 17, 1871.

[11]Benison, Saul, *Railroads, Land and Iron: A Phase in the Career of Lewis Henry Morgan* (University Microfilm, Ann Arbor, Mich., 1954), p. 312.

at since the burning of the Champion Furnace and not having been paid for months, took charge of the kiln station and would not allow charcoal to be taken from the kilns until their back pay was forthcoming. A week later the dispute was settled but some of the instigators were discharged after being paid.

From *Beard's Directory of Marquette County*, 1873.

On January 17, 1876, Donkersley, who for a while was called the "patron saint of the Morgan Iron Co.," blew the furnace out, reasoning that making iron at $23 per ton was not profitable. A month later many of the people at Morgan were destitute as they relied on the furnace for their livelihood. Although Donkersley had sympathy for the village people and thought of putting the furnace in blast, risking a personal loss, it would be many months before the furnace was operating again. "He [Donkersley] does not see the propriety of using up good material without making any profits on his metal, except to give employment to his laboring people."[12]

The Morgan furnace went back into blast in August, 1876, in a last attempt to make iron at a profit. For four months the furnace operated, but because of the continued depression of the iron market and with no prospect of an increase in the selling price of the pigs, Donkersley decided to shut the works down. The production of cord wood and charcoal was halted in the woods and the furnace ran until all the fuel on hand was consumed, finally blowing out in December of 1876, never to be used again.[13]

Though the furnace grew cold, the Morgan location itself with several fine houses standing near the furnace remained inhabited for a number of years. The Eagle Mills (then owned by F. W. Read & Company), located about one mile west on the Marquette, Houghton & Ontonagan Railroad, experienced a fire in July of 1878 which destroyed the large boarding house, sawmill and machinery, and 250,000 feet of logs — a $7,000 loss with no insurance. This was apparently the reason families of men employed at Eagle Mills were living at Morgan in June of 1881, when a brush fire moved in from the south and totally destroyed the abandoned furnace.

While the Detroit, Mackinac & Marquette Railroad was being constructed through the Morgan location and west to the iron mines in 1883 and '84, many of the railroad laborers and wood choppers of the Iron Cliffs Company lived in the houses which required major repairs after so many years of idleness.

THE PANIC OF 1874 TAKES ITS TOLL

For a time in 1871, the Greenwood Furnace consumed 125 bushels of charcoal to the ton of iron, using soft coal (softwood charcoal) exclusively, with the ore coming equally from the Washington and the Lake Superior Mines. In May, Manager Colwell switched to hard coal and production increased substantially.

In 1873, the forests surrounding both the Greenwood and the Michigan Furnace, four miles west, were leveled for charcoal. To ease the hauling of the heavy cord wood to the kilns near the Greenwood, a plank road 10 miles long was built to a battery of seven new kilns, and at Clarksburg

[12]*Mining Journal*, November 27, 1875.
[13]*Ibid.*, December 9, 1876.

another plank road was constructed leading to a set of six new kilns.

Early in 1874, 10,000 tons of pig iron were stacked at the Greenwood, and even though the orders for their product had slowed, it and the Michigan kept "blowing" steadily. The operation of these two stacks was excellent as both had been in blast almost continually since their inception. At Clarksburg it was said that ". . . The furnace keeps up a spark of life — several sparks in fact — and the cinders are all over the place . . . ," and about the railroad, ". . . the coming of the iron horse — that relentless transgressor upon the sanctities of nature — is the great excitement of the day . . . ."[14]

During November, 1874, near record-breaking production took place at both furnaces owned by the Michigan Iron Company. Ore charges made in each stack consisted of 100 pounds of Lake Superior hematite, 300 pounds of Lake Superior slate ore, 100 pounds of Shenango hematite and 300 pounds of Washington magnetic ore — totalling 800 pounds. Eighty-one of these charges were made in each furnace daily. One-hundred and twenty-two tons were made weekly at the Greenwood, and at Clarksburg, 132 tons. In 1874, the Greenwood produced 5,438 tons, and the Michigan made 7,379 tons.

Both furnaces did well over the years but the financial panic finally affected them as it had others, and on January 4, 1875, the Michigan Iron Company filed bankruptcy and the iron making concerns were turned over to H. J. Colwell, James Pickands, and A. A. Ripka to dispose of the properties and satisfy creditors.[15] The Michigan was blown out for repairs on January 12, and on April 9, the Greenwood was blown out because of exhausted ore and fuel supplies, and impossible roads. With debts, bad roads, repairs to be made and no orders for iron, neither furnace returned to service, and by 1880 the machinery was dismantled and removed.

Captain William Ward eventually purchased interests of the Michigan Iron Company property, and the sawmill at Clarksburg supported a few people who chose to remain. During most of this time, Captain Ward charged no rent for the use of his houses and the property, but in 1883, he decided to charge a small amount. The idea of paying rent unbalanced a member of the small community and in November a "fire-bug" torched the vacant iron works and destroyed the main wooden buildings, including the casting house. A week later, the sawmill was totally destroyed in a $15,000 fire, and the same arsonist was blamed. Though very little is mentioned in local papers of Greenwood after the operation of the furnace closed, Clarksburg remained fairly well populated, and sawmills owned by F. W. Read & Company and others operated there for many years, cutting pine and hardwood, much of which was used in the Marquette County area.

[14]*Mining Journal*, November 21, 1874.
[15]*Ibid.*, January 16, 1875.

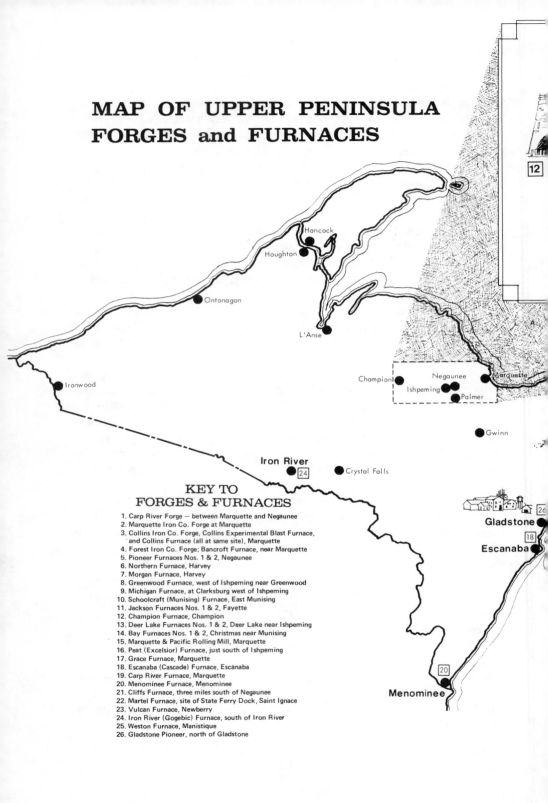

MAP OF UPPER PENINSULA
FORGES and FURNACES

Hancock

Houghton

Ontonagon

L'Anse

12

Ironwood

Champion   Negaunee   Marquette
Ishpeming
Palmer

Gwinn

Iron River   Crystal Falls
24

KEY TO
FORGES & FURNACES

1. Carp River Forge — between Marquette and Negaunee
2. Marquette Iron Co. Forge at Marquette
3. Collins Iron Co. Forge, Collins Experimental Blast Furnace, and Collins Furnace (all at same site), Marquette
4. Forest Iron Co. Forge; Bancroft Furnace, near Marquette
5. Pioneer Furnaces Nos. 1 & 2, Negaunee
6. Northern Furnace, Harvey
7. Morgan Furnace, Harvey
8. Greenwood Furnace, west of Ishpeming near Greenwood
9. Michigan Furnace, at Clarksburg west of Ishpeming
10. Schoolcraft (Munising) Furnace, East Munising
11. Jackson Furnaces Nos. 1 & 2, Fayette
12. Champion Furnace, Champion
13. Deer Lake Furnaces Nos. 1 & 2, Deer Lake near Ishpeming
14. Bay Furnaces Nos. 1 & 2, Christmas near Munising
15. Marquette & Pacific Rolling Mill, Marquette
16. Peat (Excelsior) Furnace, just south of Ishpeming
17. Grace Furnace, Marquette
18. Escanaba (Cascade) Furnace, Escánaba
19. Carp River Furnace, Marquette
20. Menominee Furnace, Menominee
21. Cliffs Furnace, three miles south of Negaunee
22. Martel Furnace, site of State Ferry Dock, Saint Ignace
23. Vulcan Furnace, Newberry
24. Iron River (Gogebic) Furnace, south of Iron River
25. Weston Furnace, Manistique
26. Gladstone Pioneer, north of Gladstone

Gladstone
26

Escanaba
18

Menominee
20

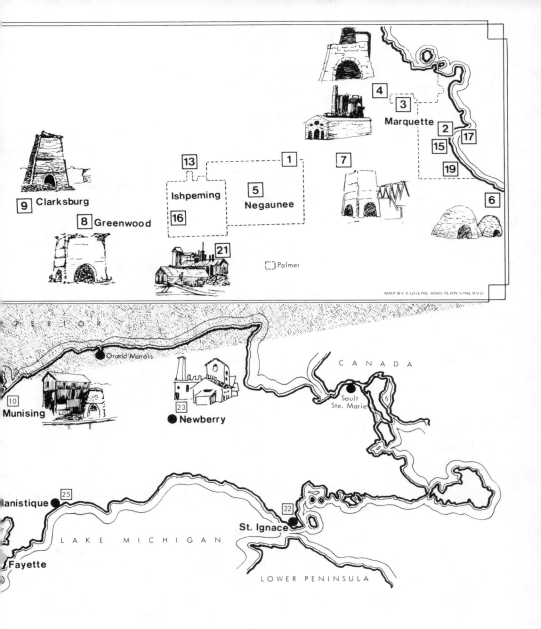

9 Clarksburg

8 Greenwood

13 Ishpeming

16

1

5 Negaunee

21

4

3

Marquette

7

2

17

15

19

6

☐ Palmer

MAP BY EUGENE AND JEAN SINERVO

P E R I O R

Grand Marais

10
Munising

23
Newberry

C A N A D A

Sault
Ste. Marie

25
anistique

Fayette

L A K E   M I C H I G A N

22
St. Ignace

L O W E R   P E N I N S U L A

Chapter VII

FIRE

Fire was a constant threat to iron-making communities. The shrill whistles of the furnace, mines or locomotives, which were used to signal fires, were frequently heard by these hardworking people. The furnace, while in blast, would ". . . emit great clouds of sparks and flaming brands . . . ," and a fire could be expected at any moment if a strict watch was not kept.[1] The furnace hands periodically wet the roofs of the surrounding buildings with hoses connected to pumps for this purpose, in an attempt to prevent the costly fires.

Besides fires started from furnaces, forest fires became more and more frequent as surrounding stands of timber were chopped down to feed charcoal kilns. The branches and tops lay on barren hillsides and flatlands of every acre cut, and in the early spring the annual forest fires endangered human lives in the outlying communities and leveled many farm buildings. Additional hazards were created when the white pine lumbering developed in the 1870s.

★ ★ ★

The Iron Cliffs Company put two new sets of kilns burning in 1867, named the Baldwin and the Excelsior, which gave the company a total of 42. The Pioneer stacks were then running exclusively on Jackson hematite ore, the favorite ore for all the furnaces when available. No other company could get the ore because the C&NW Railway was shipping it to Escanaba; the MH&O Railroad line no longer served the mine.

★ ★ ★

The rail connection with Escanaba was another of Charles T. Harvey's accomplishments. He organized the Peninsula Railroad Company in 1863, because of the longer shipping season available at Sand Point (Escanaba) on Lake Michigan. On the Fourth of July that year, he broke ground at Escanaba and track laying was started. Harvey later united with the C&NW which eventually bought his interest in the Peninsula Railroad for $50,000. On December 22, 1864, 63 miles from Escanaba to Negaunee were completed. Winter delayed the first shipment until May 12, 1865, when the Jackson Mine shipped the first iron ore over the route.

Work on a substantial dock at Escanaba was started in the fall of 1863, and in May of 1865, the first ore was dropped into its pockets. The dock was big enough to load 12 of the largest boats on the lakes at one time.

---

[1]*Iron Agitator*, October 9, 1880.

The new railroad and dock facilities at Escanaba were called by a local paper ". . . a most valuable auxiliary to the business of the Upper Peninsula." At the close of the 1865 season, 31,078 tons of ore had been moved across the Peninsula to Escanaba, and the following year the total increased to 116,271 tons.[2]

The No. 2 dock at Escanaba was built in 1872, and on October 20, 1879, work was started on a third dock which was completed on March 31, 1880.

Numbers 1 and 2 had wooden chutes; No. 3 had iron chutes and was much bigger, holding 20,000 tons of ore. Being 1,200 feet long, it had 3,620 piles to support over 4,000,000 feet of timber that went into its construction, mostly pine. All three docks were lighted in 1880 with Brush electric systems, which greatly aided the large crews of trimmers working at night.[3]

On May 11, 1880, the *S. T. Tilden* was loaded with the first ore from the No. 3 dock. In December the No. 2 dock was torn down, to be replaced with a much larger structure. In August of 1881, the first load of ore to go in an iron hulled ship at Escanaba was received by the barge *Brunswick*, which steamed out of the harbor with 1,500 tons aboard, drawing 12 feet, four inches of water.[4]

In July of 1882, the *Onoko*, newly built at the Globe Iron Works in Cleveland and the largest all-iron ship (except for decking and spars) on the lakes,[5] loaded 2,575 tons of iron ore at Escanaba. The vessel could have taken over 2,600 tons except for 70 tons of extra coal on board.[6]

★ ★ ★

The No. 2 stack of the Pioneer was blown in with a patented Lurmann's hearth in December of 1868, which increased the iron production while burning less coal. In December of the following year, No. 1 received a Kaolin hearth of the type used in Connecticut charcoal furnaces. It was six and one-half feet thick with Ohio stone on the bottom, Connecticut stone on top, and iron in between. It was designed for higher iron production and longer life.

The city fire department of Negaunee was summoned by the whistle of the switch engine *Monitor* to the flaming buildings of the Pioneer on June 21, 1877. Fire engines No. 1 and 2 responded to the alarm and had the fire under control within an hour. But by then No. 2 stack was put out of commission and both the blowing and steam engines were destroyed. No. 1 stack was not injured, but the machine shop and all the machinery was ruined. In October, the No. 1 stack was back in blast but the No. 2 stack did not get fired up until March of 1879.

---

[2]*Lake Superior Mining & Manufacturing News*, September 14, 1867.
[3]*Mining Journal*, July 31, 1880.
[4]*Ibid.*, August 20, 1881.
[5]A. O. Backett, ed., *The A B C of Iron & Steel* (Cleveland, 1925), p. 40.
[6]*Mining Journal* (Weekly Edition), July 15, 1882.

Both stacks of the Pioneer were in blast in 1879, and the iron works was in better condition than ever before. The two stacks, 49 feet high with boshes of nine and one-half feet, had bells and hoppers. No. 1 was tapped three times every 24 hours, while No. 2 made five casts in the same time. James A. Root was the founder, Edwin David chief engineer, and the pig iron made was mostly No. 1 grade, with many tons of car wheel iron being made daily. The combined average weekly product was 450 tons.

The blowing engine for No. 1 was a horizontal type, with a 22x36 inch steam cylinder powering two seven-foot blowing cylinders, while No. 2 had a McIntosh upright, with a 24x48 inch steam cylinder running a five-foot blower.

Four hundred men in all provided the labor, with 60 at the furnace, 300 chopping, coaling, and mining, and the rest doing miscellaneous work. Annually, the furnace burned 2,000,000 bushels of charcoal, and 40,000 tons of ore ran through her two crushers. The ore was mined entirely from the Jackson Mine, lying west in the city, and was hauled to the furnace in wagons. Nine kiln locations were kept busy, having a total of 60 kilns, some on rail lines, and others deep in the forests, and every hour of the day wagons or rail cars would deliver charcoal to the furnace.[7]

The founder strove to keep the furnace in blast and continually watched the hot blast temperature and pressure, the condition of the slag run out, and a hundred other things to assure long life to the hearth, which was most effected by varying temperatures. An 18-month run was had by the Pioneer No. 2 on one hearth, ending on January 8, 1881, after making 14,652 tons of pig iron.

James A. Root, 13-year veteran of the Iron Cliffs Company working as founder at the Pioneer furnace, began an independent charcoal business near McMillan in Luce County in August of 1885, with his son-in-law, D. L. West. Both were from Negaunee. Their holdings included 3,000 acres of hardwood lands on the Detroit, Mackinac & Marquette Railroad, and charcoal was made in a test set of six of Lovelace's iron kilns, which produced an excellent grade of charcoal. But the business was short lived as Root died seven months later in a rail accident in the Pioneer yards.[8]

Fire struck in the Pioneer Furnace engine room at 1:15 a.m. on September 21, 1887, doing $15,000 in damages before being brought under control. It was thought that one of the night workers went into the brick veneered building where barrels of oil were stored to refill his lamp and, setting his lighted ''torch'' on a barrel, had ignited this inferno. The engineer on duty was able to give only five short blasts of the whistle before fire forced him out, but luckily the night engineer at the Lucy Mine spotted the blaze and blew his steam whistle to awaken the town. The engineer was unable to shut off the steam to the blowing engine and all the

---

[7]*Mining Journal,* August 16, 1879.

[8]Called the Lovelace Iron Charcoal Kiln, the patent was issued on October 6, 1885, and jointly owned by Walter A. Lovelace and Dudley G. Stone, of Negaunee.

while the Negaunee firemen laid out hose (nearly 3,000 feet), its thumping could be heard. The machine shop and an old engine room also caught fire and lathes, pumps and other machinery were extensively damaged. While the fire burned the wooden interior of the engine room, the windows fell out and the great flywheel on the Corliss engine could be seen spinning loyally, the engine yet unhurt.

The iron in the stack was all run out normally, thus avoiding a salamander, and the engine was found to be in fair shape after the fire, needing very few repairs to be run again. In the first week of October, the furnace was again making iron. The loss was covered by the Peter White & Company insurance agency in Marquette which held a $40,000 policy on the iron works.[9]

The first electric lights in the city of Negaunee were turned on at the Pioneer Furnace on March 14, 1889, powered by a machine installed by the Western Electric Company of Chicago. Arc lights were hung throughout the plant which were said to be a "great assistance" to the night crews, and at the end of a 30-day period on May 18, 2,570 tons of iron were made in the No. 1 stack (No. 2 having been out of blast for some time). The old 57-foot-high No. 1 stack with a bosh of 10 feet had turned out 92 tons in one day during this period, which was the largest day's production for a charcoal furnace in the United States except for the new Ashland, Wisconsin, furnace which produced 100 tons daily.

The stack at this time had been in blast for 28½ months (except for a three-week shut down caused by a fire) and had produced 52,000 tons of charcoal pig iron. It remained in blast until October of 1889. The lining by then was badly burned and the No. 2 stack was put into blast in mid-October after it was completely relined with fire brick.

REBUILDING THE ISHPEMING PEAT FURNACE

Hiram A. Burt of the Carp River Iron Company leased the Peat Furnace in September of 1879 with the 10 Kloman kilns, and contracted for six kilns to be built at Barkville with an additional six to be built at Spalding on the C&NW Railroad, a long haul from the iron works. The blowing engine, boilers and other machinery that had been used at the Greenwood were purchased from the assignee of the Michigan Iron Company, as the Peat Furnace machinery had been stripped out during the furnace's idle years, and put to use in the company mines.[10] The casting house was rebuilt measuring 35x53 feet, a new stock house of 80x100 feet was erected, and an elevator installed to do away with hauling the stock up the steep hill with horse and wagons.

With the charcoal being made elsewhere, the three kilns located at the furnace were converted into a tool house, store house, and a blacksmith

[9]*Mining Journal* (Weekly Edition), September 24, 1887.
[10]*Ibid.*, September 13, 1879.

shop. In December, the Peat Furnace was renamed the Excelsior, and it retained this name until its last blast.

THE EXCELSIOR

Many improvements were made, one of which was the changing of the furnace from an open front to a closed front, leaving only two notches for iron and cinder to run out. Finally, on April 4, 1880, it was put in blast and the next day three and one-half tons were cast. A month later the furnace was making 24 tons a day. Ore used was teamed from the nearby Lake Angeline Mine, and the blast was started with 20,000 bushels of coal on hand.

At 2:00 a.m. on Wednesday, June 9, 1880, mine whistles, steam whistles, and the fire bell awoke the people of Ishpeming, in time to see flames and smoke rising into the night from behind Burt Hill as the Excelsior Iron Works was destroyed. A charge hanging in the stack had suddenly dropped and forced flames and a shower of sparks through the flues, which had started fires in many sections of the furnace.[11] All the improvements so recently made were destroyed as well as the newly purchased engines and blowers. This fire reduced the number of stacks in the Upper Peninsula to 16, and only six were in blast.

The new Knowles fire-pump system had failed, and Chief Engineer Anthony, fearing an explosion of the boilers, had opened valves to vent the rapidly built-up steam pressure, burning himself badly in the process. Two weeks after the fire, 60,000 bricks were delivered to the iron works for rebuilding, and late that fall it was back in blast. Iron was then made quite steadily until 1883, when the depressed iron market forced the works into another period of idleness.

When the Excelsior Furnace shut down in February, 1883, 40 men were thrown out of work with back wages for four months still owed them. Their families suffered greatly that winter as the company could not pay the $10,000 owed in back wages, and many of the households ran out of both food and fuel depending on what their friends could provide for survival.

MARQUETTE AND PACIFIC

The Carp River Iron Company, unable to fill charcoal iron orders because of the Excelsior fire in 1880, decided to turn its Marquette & Pacific Furnace into a charcoal stack in order to use the fuel that was rapidly accumulating at their kilns. During the change-over, a large salamander was found in the bottom of the stack, which would not have mattered if bituminous coal and coke had been used again. But with the change to

---

[11]*Iron Agitator*, July 17, 1880.

charcoal it had to come out. To accomplish this, a few charges of black powder were applied, breaking and loosening it enough for removal.[12] Then, in September of 1880, with a new bell and hopper manufactured by the Iron Bay Foundry added to the stack, the furnace went back into blast. Patrick Carroll was the new founder and under his care the stack produced as much as 45 tons a day.

The Marquette and Pacific remained in blast just long enough to meet the company's charcoal iron commitments and the last reference found of it being in blast, was in August of 1881.[13] In April of 1882, there were rumors to the effect that it was going to be changed to a coke iron furnace, but this was later denied and they simply readied it for emergency use only. For the next 20 years the sandstone buildings stood quiet and empty most of the time and they never heard the hissing and roaring of the steam engines or felt the warmth of the furnace again. The old iron-shelled stack, boilers and machinery were torn out and sold for scrap late in 1899.

Over the years, the walls of the buildings weakened and cracked, until finally, in June, 1901, a ruinous storm struck the city with high winds and driving rain. The buildings collapsed amidst clouds of dust, killing one Chet Michaljohn and four horses, and severely injuring seven other horses. Thirty-three-year-old Michaljohn was a new member of a band of horsetrading Gypsies. The Gypsies had stabled their horses in the Marquette & Pacific buildings without the permission of Peter White, the owner, and Michaljohn and two young boys were inside tending them when the storm hit. Hearing the walls crack, the three made for the door. Two boys escaped in time but Michaljohn and the horses were buried in timbers and stone.[14]

THE GRACE FURNACE

The Grace Furnace, idle on the Marquette harborfront since 1874, was purchased by the Lake Shore Engine Works in 1902 and in October of that year dismantling was started. After the machinery was removed, the exterior sandstone walls were taken down and all the iron trusses and braces were removed for possible future use. What was left of the buildings was demolished and the stack was brought down by pulling out the supporting columns.

Bishop J. G. Pinton of the Marquette Catholic Diocese, "closed a deal" with the Lake Shore Company on April 1, 1903, and acquired all of the sandstone building material which had been salvaged from the old furnace. Originally from the Marquette quarry, the stone closely resembled that used in St. Peter's Cathedral, and was used by the parish to build

---

[12]*Mining Journal*, July 17, 1880.

[13]*Ibid.*, August 27, 1881.

[14]*Ibid.*, (Weekly Edition), June 28, 1901.

Grace Furnace, Marquette, with ore dock and sail boats in background.

Bishop Baraga High School, which stood on the corner of Superior (Baraga Ave.) and Fourth Streets for almost three quarters of a century.[15]

The city of Marquette purchased Baraga Central High School and razed the mammoth building in 1974. Some of the sandstone was salvaged from the old school and has since been incorporated in the decorative outside walls of the new city hall building, which occupies the old high school site. These sandstone walls and the retaining walls of the new city hall, built with stone used originally in the construction of the Grace Furnace, are a fitting tribute to and reminder of our blast furnace history.

FIRE DESTROYS THE CARP RIVER FURNACE

William Beals, founder of the Carp River Furnace, was forced to bank the furnace many times in 1879 and 1880, because of shortages of charcoal at the company's kilns, and the Chocolay farmers' refusal to make charcoal at nine cents per bushel. They claimed it cost them $1.00 a cord to have the wood cut which left no room for a profit on the coal.

The hoisting house of the Carp Furnace was destroyed by fire in August of 1880, but with a horse-powered lift set up to raise the ingredients to the top of the stack, the furnace was back in blast 30 hours later. On the

[15]*Mining Journal*, April 2, 1903.

afternoon of November 17, 1882, the engine house caught fire and burned unchecked, quickly spreading to the stock house. The smoke rising from behind the high hills south of Marquette drew many spectators and while they watched four ore cars and one box car were destroyed, along with the stock house, top house, floor of the tankhouse, and the water tank. The steam pump was used to fight the fire for a short time, but somehow the steam valve of the engine opened and with the loss of pressure the pump came to a stop. A last futile attempt to control the fire with a bucket brigade failed and the fire soon raged unrestrained, doing damage which stopped iron making at the furnace until 1890.

THE MARTEL AND VULCAN

The opening of lands lying east of Marquette to Point St. Ignace in Mackinac County by the building of the Detroit, Mackinac & Marquette Railroad added years of life to Upper Peninsula blast furnaces, whose hardwood lands were rapidly being cut over. Immense stands of hardwoods, untouched by furnace colliers, waited for this railroad in Schoolcraft, Chippewa, and Mackinac Counties.

As soon as the first 20 miles of track were laid east from Marquette, John Burt organized the Union Fuel Company, and negotiated for large timber holdings along the line. The railroad also prompted building of blast furnaces at St. Ignace and Newberry.

Although the railroad was talked about for many years, no action was taken until October of 1879 when crews of surveyors with three teams of horses loaded with provisions left Marquette to mark the most desirable route. The first 20 miles were let to a Canadian railroad building firm, McDermott & Hendrie. On January 15, 1880, men and teams arrived in Marquette to start work on the line. The last spike for the first 20 miles was driven by the editor of the *Mining Journal* on July 25, 1880, and by the following July, tracks had been laid to a point 10 miles southeast of Munising. Track laying was also started westward from St. Ignace.

The Union Fuel Company built 11 kilns on the south bank of the Carp River near the furnace and batteries of kilns at Whitefish, Glenwood and Rock River. The Glenwood kilns were burning in January, 1881.

Many cedar swamps and tamarack bogs are on this route of the railroad and mosquitos swarmed on the crews of men, testing tempers and making them wish they had never left their homeland. Most of the workers were Italians, Swedes and Canadians who found it hard to tolerate each other and fights were not uncommon. Several of the Swede laborers were admitted to the county poor house at Marquette in August, 1881, suffering from malaria which they contracted while working on the DM&M.[16]

---

[16]*Mining Journal*, (Weekly Edition), September 3, 1881.

Martel Furnace, St. Ignace, 1903.

The road extended 92 miles east from Marquette by November, 1881, and 50 miles west from St. Ignace to just past Newberry, which was named after ex-congressman John S. Newberry, one of the railroad's directors. The two rail ends were joined a short distance west of Newberry near Dollarville on December 9, 1881. The Hon. Peter White was asked to participate in the ceremonies, and after a short speech in which he said, ". . . I now drive the last spike, to join the upper peninsula to the lower . . . ," he hammered in the spike.[17]

Work on the St. Ignace ore pier was started in late November of 1881, and 10,000 lineal feet of piles were driven to support the structure which was to hold 5,000 tons of ore when completed in August of 1882. The pier was 675 feet long with 50 pockets on each side, had a shore approach 550 feet long, and was able to load six vessels at the same time. In early May of 1882, the pier (not yet completed) was loaded with ore for the first time from Marquette County iron mines, and the first St. Ignace ore shipping season was underway.

Davenport & Fairbairn, owners of a large railroad car wheel works at Erie, Pennsylvania, began construction on the Martel Furnace at St. Ignace in the fall of 1880 in order to make the high grade charcoal iron they

[17]*Mining Jounral* (Weekly Edition), December 17, 1881.

preferred. The stack was an iron jacket, 53 feet high, with bosh of 10½ feet, and was located on a 10-acre site on the bay. The casting house and boiler house were made of brick, and measured 43x70 feet and 25x36 feet, respectively. Four boilers, each 32 feet long, generated steam for the massive blowing engine which was housed in a building 26x26 feet. Flux was taken from a limestone ledge one quarter mile from the works, and 24 kilns of 50-cord capacity were constructed for coaling, drawing wood from 15,000 acres owned by the company and situated on the DM&M Railroad.

The *Mining Journal* of September 10, 1881, carried an interesting article relating to the furnace which was then under construction:

> The story of Sau-ge-mau's relentless massacre of the Iroquois who lived in a village on extremity of Point St. Ignace, which occurred some 225 years ago, is recalled . . . by the finding of a great quantity of human bones in a sandy ridge a few rods beyond F. R. Hulbert's fine residence. The ridge was opened for the purpose of getting sand for use at the furnace, and the bones were found in heaps, showing that numerous bodies had been rolled into a common grave. The village numbered 200 inhabitants, and but 25 escaped the fury of Sau-ge-mau. These fled to the islands, secreted themselves in Skull Cave, and there miserably died. Bones of both sexes, and of children as well as adults, have been found. Over the huge grave had grown up a pine of nearly a foot diameter, which the march of improvement cut down not very long ago. Tourists have carried off most of the skulls, those of children being preferred by our singular guests. Nearly every skull was crushed in on one side as if done by a blow from a tomahawk.

The Martel Furnace was blown in on August 14, 1881, using ore which was shipped to the furnace by boat, and during the 106 days of operation that year 4,109 tons of metal were made, averaging about 39 tons per day. The furnace, like the old Deer Lake No. 1, was also banked on Sundays. During the week of December 24, 1881, the Martel Furnace received the first shipment of iron ore to pass over the new railroad line from Marquette, consisting of 17 cars of Dalliba ore, from the Dalliba Iron Company's mine located at Champion. *Charles Martel is a distant relative*

It was said that the Martel was named after the French King who stopped the invasion of Christian Europe by the Saracens. The owners of this furnace somehow believed it would put vice and intemperance on the run, and provide "sturdy blows for the right." How a blast furnace could be so oriented towards temperance is difficult to understand.

In its first year of production, the Martel Furnace produced 11,283 tons of pig iron, averaging just 87 bushels of charcoal to the ton. At this time the company was installing retorts in the wood kilns in an experiment to extract chemicals from the wood smoke during the charring process at a cost of nearly $100,000. This was the first use of retorts for coaling wood in the Upper Peninsula, and all furnaces which later went into blast — except for the Iron River — were constructed with chemical plants as an integral part of the iron works.

The company lit the fire of the first set of retorts in October, 1882, and a week later the second set was put into use. These Mathieu-styled retorts were intended to replace all of its kilns, which would be abandoned, and provide a better quality of charcoal. More retorts were added and the furnace ran smoothly for a while, producing an average of 60 tons daily, with a one day high of 73 tons. Another depression in the iron market closed the works down in March, 1883.

The chemical works of the Martel Iron Works included a sawmill which was started up on August 1, 1883. The following day, fires were lit in the Whitewell hot blast stoves, in preparation for the furnace to go into blast. The purpose of the sawmill was to cut hardwood logs into a desirable size for making charcoal. The logs were first cut into three-inch planks, then moved by live rollers to a trimming table. Here, they were caught and fed into "button" saws, positioned from two to three feet apart. The pieces were then moved on a carrier and dropped into the wood storage house.[18]

Fifty-six retorts were put into operation, eight at a time called a "bench," and the furnace was started and run for a few days on kiln-made charcoal. The chemical works concentrated on making wood alcohol and acetate of lime, but probably little was produced as the furnace was blown out again in late November.

VULCAN FURNACE

The Detroit corporation which had built the DM&M railroad, started work on a large blast furnace at Newberry in the spring of 1882, under the name of the Vulcan Furnace Company. James McMillan was president and other officers were John S. Newberry, Hugh McMillan, F. E. Driggs, Lee Burt, and Charles A. Burt. Having controlling interest in the railroad gave them the advantages of cheaper rates for hauling both iron ore and pig iron, which allowed the furnace to operate at times when the iron markets were unfavorable, and even when the Martel Furnace was shut down, unable to make a profit.

Newberry Furnace was only a few months old by July but already 300 people lived there. Among them was Dr. C. H. Voorhis, who had resigned as the physician of the Union Fuel Company at Onota and taken up his work with the Vulcan people. The furnace was well underway by then, many houses were up, and the company store was in business. Retorts for the Vulcan chemical plant were landed in St. Ignace by the car ferry *Algomah* in August, 1882, and were sent on to Newberry. It was their intention to run entirely on charcoal made in these retorts.

---

[18]*Daily Mining Journal*, August 4, 1883.

The Vulcan Furnace, supplied with charcoal from 32 of the 56 installed retorts, went into blast on May 21, 1883, casting its first iron on the next day. The retorts worked well for a time, furnishing the charcoal needed, and the first shipment of wood alcohol was made to a Detroit firm. A fire broke out in the plant on July 15 of that year and before it was subdued, the retorts, sawmill, and the chemical works had suffered $100,000 in damages, though the furnace was untouched and stayed in blast. The Union Fuel Company then supplied what charcoal it had in stock, but this was not enough to keep the furnace in blast and it was shut down that fall.

The company rebuilt the chemical plant using brick walls with galvanized iron roofs, and in November, 1883, the retorts were again fired up and the furnace went back into blast. The production of the Vulcan Furnace for 1883 was 4,158 tons of pig iron, while the Martel Furnace made 5,747 tons.

The Farm Mine located in Michigamme supplied eight cars of iron ore daily to the furnace and in January of 1884, W. L. Wetmore of Marquette and Charles H. Schaffer of Onota were supplementing the retorts with charcoal made in their kilns. In September, an electric light plant was installed at the furnace which supported a number of Edison 16-candle power lights placed throughout the iron works. The furnace output was averaging 60 tons per day while the retorts produced 200 gallons of alcohol, but the retorts were still not producing a sufficient amount of charcoal.[19] The Vulcan Furnace was described as doing "magnificent work" in May of 1884 when it was throwing off a blast every eight hours, which averaged about 10 beds. Each bed apparently had enough moulds formed in it to contain two tons of molten metal. The Vulcan Furnace produced 11,080 tons during 1884.

Forest fires occurred frequently in the denuded hardwood lands surrounding the kiln stations and settlements, causing economic setbacks to the already hardpressed companies and the Vulcan Company was no exception, having had more than its share. In 1884, 3,000 cords of its valuable wood were destroyed in a fire that encircled the small village. Five thousand cords of stacked wood and seven of the iron company's story-and-a-half cottages were razed by the flames in 1885. The following year, 7,000 cords valued at $15,000 were burned in an early spring fire.

The May 30, 1885, issue of the *Mining Journal* reported of the Vulcan Furnace: ". . . sweeping changes are under way (which) will relieve the plant of the costly and ineffective system of retorts, and the chemical works . . . have proved a discouraging failure. (They) will substitute kilns for the retorts . . . the good old way . . . ."

The furnace company also pushed its narrow gauge railroad a number of miles out into the hardwoods, putting a strain on the little pony engine *Vulcan* and forcing it into the shops for an overhaul. Forty-two kilns were

---

[19]*Iron Agitator*, June 14, 1884.

in operation in November of 1885 and the following month the repaired *Vulcan* was on the run again, hauling from deep in the woods to the kiln station up to 150 cords a day. In May of 1886, seven more kilns were added for the coaling operation. The furnace was then receiving its ore from the Chelsea, Detroit and Wetmore Mines, and had turned out 26,000 tons of iron in an 18-month blast ending in February of 1887.

Fire again struck the Vulcan Furnace in June of 1886, when a DM&M locomotive pulling cars into the iron works set off a fire which completely destroyed the ore storage sheds and two ore cars.

Charles A. Burt resigned as superintendent of the Vulcan Furnace in April of 1883 and left the area. But in the fall of 1888 he moved back to Newberry from Detroit and took the same position on the retirement of Royal A. Jenny. Plans were developed at what was then called the Newberry Furnace to again make chemicals from the charring hardwood and 100,000 brick were purchased from the Anna River Brick Company of Munising to erect an acid works.

The furnace operated sporadically through the years and in April of 1893 it was making up to 80 tons per day, but in 1894 the furnace and chemical plant were shut down.

In 1897, there were doubts the furnace would ever run again and reports indicate the chemical plant was being torn down, to be shipped to the Weston Furnace in Manistique where it was to be assembled to make wood alcohol.[20]

Furnace at Newberry, showing gondolas of pig iron.

The Newberry Furnace sat idle for eight years and in October of 1902, it was purchased by the Michigan Iron Company, Ltd. Work was immediately started to renovate the iron works. All of the old machinery, including the hot blast ovens, were removed, and kilns in serious dis-

[20]*Mining Journal* (Weekly Edition), October 9, 1897

repair were demolished. Forty new kilns of 80-cord capacity were built by C. H. Peterson of Traverse City, and a railroad five miles long was put in, reaching to 16,000 acres of newly-purchased hardwood lands north of Newberry.

The furnace, when completed, was equal to a new plant and, in March of 1903, it was put into blast. The stack was then 50 feet high with a bosh 10½ feet, and the 800 degree Fahrenheit blast was blown in at five and one-half pounds pressure. For the month of April, 1904, the furnace produced 80 tons of pig iron per day, making in one day 87 tons.[21] Another large chemical plant was added shortly after starting up, and the concern went on to produce iron and wood alcohol and other chemicals for many years after.

THE IRON RIVER FURNACE COMPANY

The 28th stack to be built in the Upper Peninsula was that of the Iron River Furnace Company near Stambaugh, 700 feet from the entrance of the Nanaimo Mine. John S. McDonald was president of the company; Alexander McDonald, vice president; Louis Muenter, treasurer; John Spence, secretary, and John T. Jones, manager. The capital stock was a mere $50,000, with 500 shares at $100 each.

In November, 1883, the company purchased the machinery which survived the Bay Furnace fire. It consisted of two engines, four boilers, and a hoist, all for $1,900.[22] The first work, started in December of 1883, was that of clearing off 20 acres for the iron works, and excavating for the kilns. Ten of the 30 planned kilns were completed in March, and 50 men were chopping cord wood, to be ready for the day the furnace was blown in. By the time the first iron was made, this wood would be rotting in heaps. The Bay Furnace machinery arrived in Ishpeming on the DM&M in January, 1884, and was sent on to Iron River.

Construction at the site began well and the 80 to 100 men employed soon had a store, boarding house, barn and office with a bridge spanning the river. Construction was also started on the furnace buildings but plans were larger than financial resources and in July, 1884, the Fond du Lac and Iron Mountain capitalists went into bankruptcy.

Very little was done at the furnace until the company straightened out its financial affairs later that year. In March of 1885, 40 men were at work positioning the machinery, working on the buildings and filling and burning some of the kilns. The stack was 56 feet high with boshes of 11 feet and the hearth measured six feet by eight feet. The two patented Merrill engines ran the blowers, and hoisting was done with two 48-inch drums. On February 2, 1886, the furnace made its first cast and in March it was producing 40 tons per day, supplied largely with coal from the 18 kilns,

[21]*Daily Mining Journal*, May 5, 1904.
[22]*Mining Journal* (Weekly Edition), November 17, 1883.

the remainder being shipped in by Charles Schaffer. The company was also engaged in casting 1,200 iron plates, needed in the construction of W. A. Lovelace's patented portable kilns, eight of which were in running order in July.[23]

The ore for the furnace was provided from the Nanaimo Mine by way of a long trestle and an endless rope 1,750 feet long. After the skip was loaded with ore in the deep mine, it was hoisted to the surface and dumped into a small tram car, pulled by the long rope on the trestle and dumped into a pocket where the crushers were located. From there it passed into a calciner. The main difference in the iron making process at the Iron River Furnace was the use of the calciner which heated the ore nearly red hot to remove all of the moisture. Another recorded instance of heating the ore to remove the moisture was at the Carp River Forge where a masonry kiln was used. The type of construction of the Iron River Furnace calciner is unknown. The crushed and dried ore, called "sponge ore," when removed from the calciner, was again placed in a skip, hauled up an incline track to the top of the stack, and dumped for charging.[24]

About the first of August, 1886, Charles Himrod, famed for his knowledge of iron making, leased the faltering iron works. It was expected that he would eliminate the debt and establish the enterprise on a solid financial footing. But fire struck in mid-July and the dreams of the struggling families were shattered. The fires started when a weakened spot on the stack gave out and fire roared through the hole, followed by a gas explosion, which threw the burning charges in all directions. The wooden stack house, dry as tinder inside from the constant heat of the furnace, burned to ashes with the ore pockets and crusher house. The *Mining Journal* for August 28, 1886, stated, "Ever since the burning of the furnace there has been a steady exodus of families from here and other parts, as everything seemed to be getting out of shape here, and nobody knew where to look for a day's work . . . ."

George C. Reis, owner of a large farm at Eden, Minnesota, purchased controlling interest of the Iron River Furnace Company in September of 1886. He intended to make repairs and put it in blast when he sold his 40,000 bushel wheat crop, but months elapsed before a sale could be made.

GOGEBIC FURNACE

The name of the Iron River Furnace was changed to the Gogebic Furnace Company in the spring of 1887 and on May 15, after a great deal of work, it was returned to blast. The furnace produced as much as 56 tons per day at times, but 50 tons was the average. The company then owned and had in operation 28 charcoal kilns; 10 at the furnace called the "Home Kilns;" the eight iron kilns located two miles from the

[23]*Mining Journal* (Weekly Edition), July 10, 1886.
[24]*Ibid.*, (Weekly Edition), November 21, 1885.

furnace, and the "East Kilns" three and one-half miles out in the hard-woods. J. A. Wagg of Green Bay had the contract to fire the iron kilns; Fred Olson with John McCall, a collier, had the East Kilns, and C. A. McRae was in charge of the entire wood cutting operation.

Wood cutting continued until April of 1888, at which time the company discharged all of its woodsmen and operated the furnace into May when it ran out of both fuel and ore and shut down. The Home Kilns were filled and produced charcoal for a while, with the charcoal shipped to Ashland, Wisconsin, where a large furnace had recently gone into blast. Thousands of cords of hardwood were left out in the slashings after the company failed and a greater portion of this was destroyed by forest fires. A local paper reported in September of 1888 that the company was ". . . so sick that it was tottering on its last legs."[25]

Shortly after the furnace was stopped, the Nanimo Mine also went into bankruptcy and hoisted out its pumps, allowing the mine to fill with water. The Iron River Furnace was never given a fair trial on good ore and its owners suffered continually from a lack of operating capital and persistent shortages of fuel. The financial collapse of the furnace forced the company to demolish its cast iron kilns and sell them for scrap in July of 1889 to satisfy labor claims placed against the furnace.

THE FAYETTE FURNACE CLOSED DOWN

Both stacks at Fayette were blown out in June of 1881 for changes and repairs. They were then raised to 53 feet and a new hot blast was installed. Bell and hopper assemblies were also added to each at a total cost of $30,000. Their production for 1882 was 8,657 tons of iron and the furnace continued to cast. However, on the evening of May 12, 1883, a fire started in the stock house and nearly destroyed the entire iron making concern.

The warehouse clerk was the first to notice the flames at the corner of the building nearest the bay and he gave the alarm. Soon, every able-bodied person was fighting the wind-driven fire. The fire engine was placed near the buildings and while ". . . men, women and children . . . worked with all their strength . . . pumping the hand engine . . . ," the inadequate stream of water was fed to the flames.[26] Those not pumping the engine carried pails of water from nearby Lake Michigan and threw them uselessly on the fire. It was thought that the whole town was doomed. Then, the steamer *Lady Washington* moved from her moorings and, positioned by Captain John Colwell, started her pumps. In a short time the fire was extinguished as it began to threaten the second casting house.

The first flames spotted by the clerk had spread rapidly inside the stock

---

[25]*Mining Journal* (Weekly Edition), September 8, 1888.
[26]*Ibid.*, May 19, 1883

house where the timbers and beams were coated with charcoal dust collected over the years. The dust acted like fast fuses and sent trails of fire in all directions. Sixty-five thousand bushels of charcoal stored there caught fire and destroyed the building. The engine house, top house and other buildings were also destroyed, amounting to a $40,000 loss. The *Lady Washington* aided in saving the two bare stacks, the engines, one casting house and the hot blasts and boilers. The steamer also saved the coal kilns near the bay, a large supply of cord wood, and the lower docks.[27]

After this setback, and with the growing problem of securing fuel constantly nagging at them, the Jackson Company seriously considered closing the iron works down, but in June the decision was made to rebuild. Captain Harry Merry of Negaunee was brought in to supervise the reconstruction. The No. 1 stack was again making iron in September, 1883.

Captain Merry was well known in the Peninsula, having been mining captain for many years. In the early days of the Jackson Iron Company, Merry was captain of one of its openings at Negaunee. The uprooted tree, under which iron ore was first discovered on Lake Superior, lay near pit No. 2 of the Jackson Mine, and it was Captain Merry who had this tree cut up, hauled to the Jackson mill and sawed into lumber — some of it used to make the furniture in his office.

The "historic stump" rested near the North Jackson Mine for many years in a fair state of preservation but a fire of undetermined origin destroyed it in 1901.

Mining captains were highly respected people in their communities, having fine houses and were usually well-traveled. Most were well treated in the local papers. Their word was law on the mining properties, underground, on the surface or in the many open pits, and the companies often named their open pits after them.

One day in January of 1885, while Captain Merry was in a saloon in Escanaba, a big fellow walked in and started pushing the barkeeper around for some past grievance, and threatening everyone in the place. Captain Merry, not used to this type of abuse, calmly drew his pistol and shot the troublemaker in the foot, putting him out of commission for a few days and restoring order as a captain must.[28] He was applauded for his community spirit.

The Fayette Furnace produced 16,875 tons of pig iron during 1884, making at times over 50 tons per day with one stack in blast. The furnace was blown out about July 1, 1885, for an "indefinite period" as the company had 22,000 tons of iron in the yards, and more in Cleveland, unsold. One stack was back in blast in March, 1886, but the scarcity of fuel and low iron prices prompted the company to consider the possibilities of either moving the furnace to Escanaba or Negaunee or shutting it down

[27]*Mining Journal*, May 19, 1883.
[28]*Ibid.*, January 31, 1885.

permanently. In 1887-88, the company had wood cutters out in the forests 15 to 20 miles from the iron works. The wood had to be hauled to the furnace by wagon over bad roads.

The furnace was banked for lack of fuel in the spring of 1887, but the following month the hot blast was turned back on. Fuel shortages cropped up on other occasions and, in early December of 1890, the Fayette Furnace ended its last blast. The *Mining Journal* of December 13, 1890, stated, ". . . Fayette closed down for the season (and there is) rumor that suspension is permanent . . . ." Later that month the paper reported, ". . . there is no money to be made by its operation." Immediately after going out of blast, the Jackson Company began moving equipment to Negaunee, but life was gone from the old machinery and the company never did try to resume operations.[29] This signaled the end of one of the most prosperous charcoal iron furnaces in Michigan which had produced in its lifetime over a quarter million tons of pig iron.

THE MARTEL FURNACE MOVED FROM ST. IGNACE

The retorts at the Martel Furnace provided a very poor quality of charcoal, far below that produced by a kiln. The company-owned kilns along the DM&M railroad burned constantly while iron was being made to meet the demand of the furnace. Six more retorts than what Mathieau had recommended were built but they did not work well and very little alcohol was extracted through the carbonization process. The company considered the retorts a failure and called ". . . the system — so perfect on paper — a fraud when materialized in brick, mortar, and iron."[30]

The furnace went into blast again on July 19, 1886, and did extremely well, making 8,450 tons by January 8, 1887, when it blew out for repairs. Other blasts were started — one in April of 1890 and another in May of 1892. But the company's financial affairs were growing worse, and on August 29, 1894, the furnace and all real property were sold to Galbraith & Plummer Brothers. The *Mining Journal* stated in the September 8, 1894, issue that ". . . This finally disposed of the Martel Furnace Company as a business concern . . . the Martel will remain idle for an indeterminate interval."

Hopes were high for reviving the St. Ignace iron industry in August, 1902, when Frank B. Baird of Marquette purchased the 21-year-old furnace for $28,500 and began to repair the long-idle machinery. Before the end of the year, the M. A. Hanna Furnace Company, of which United States Senator Mark Hanna was an officer, leased the Martel plant and put it in top condition. In January of 1903, after being idle for over 10 years, the furnace was again making charcoal iron. However, in April of that year, fire destroyed the ore stockhouse, a supply of charcoal, and

---

[29]*Mining Journal* (Weekly Edition), December 13, 1890.
[30]*Mining Journal*, September 13, 1884.

a railroad trestle. Shortly after, the Hanna Company gave up its lease on the property.

Baird and a Mr. Bolten purchased a hardwood timber stand near Ozark Lake and planned to use the Ozark kilns to make charcoal, but the St. Ignace furnace was never fired again. Baird, owner of the East New York Mine in Ishpeming, next organized the Boyne City Charcoal Iron Company and made plans to move the furnace to Boyne City in Charlevoix County. Charles Schaffer was president of the new company, Baird was vice-president, and Noah W. Gray, secretary and treasurer. Their intentions were to receive charcoal from the Boyne City Chemical Company as the supply of hardwoods was nearly depleted at St. Ignace.[31]

A crew of 25 men were busy in November of 1903, dismantling the furnace and removing the machinery for shipment to Boyne City; even the brickwork was taken down. The superintendent of the iron company was Fred Smith of Marquette, who formerly held the same position at the Carp and Excelsior furnaces, which had a large production under his leadership. In December of 1904, the furnace was up and ready to go into blast at Boyne City. It was considered the finest blast furnace in Lower Michigan.

---

[31]*Ibid.*, October 17, 1903.

Chapter VIII

THE ALGER COUNTY CHARCOAL KING

The Union Fuel Company had financial troubles in 1884 and, as a result, Charles Schaffer leased the charcoal-making concern, keeping its kilns burning into the 1900s. In the late spring of 1886, Schaffer purchased all of the timber lands and kilns of the defunct Union Company, including the Whitefish, Rock River and Glenwood kilns, and five kiln stations on the C&NW Railroad.[1]

After the Bay Furnace village of Onota was destroyed by fire in 1877, the name of Onota was applied to a location approximately 20 miles to the west, believed to be Glenwood Station, where John Burt had erected about eight kilns in 1880. Dr. W. J. Palmer lived there with his daughter, and managed the charcoal business for Schaffer, while Lachie McKennon ran the Onota store. Robert Dilinger, a well known collier, and John Shea burned the kilns at Rock River and Onota, and Mitchell Berrie was the foreman.

Schaffer had a contract with the Deer Lake Furnace Company to supply 100,000 bushels of charcoal per month during 1885, and later had contracts with the Iron River, the Vulcan, and other furnaces. To supply the enormous amount of charcoal used annually, Schaffer's crews cut thousands of cords of wood each year. In December, 1886, alone, 2,300 cords were cut (with the cross-cut saw) averaging 100 cords for every working day of the month.

Schaffer, called ". . . The Alger County charcoal king . . . ,"[2] sold half-interest in his kilns at Whitefish, Rock River and Onota to J. W. Belknap in 1890. The fuel company was then known as Schaffer & Belknap.[3] More kilns were planned for the expanding business, as the Carp River Furnace also received their charcoal. Five more kilns were built at Onota, three went in at Rock River, and three at Whitefish, for a total of 35 at the three locations. In April, 1891, articles of incorporation for the Onota Chemical Company were drawn up by Schaffer and others from Marquette. The firm planned to make charcoal and process the smoke for alcohol from the 40 kilns then in operation, but the declining charcoal iron industry discouraged the owners, and their plans never materialized.

THE SALE OF THE DEER LAKE FURNACE DAM

Residents of Ishpeming became concerned about the disposal of the

---

[1]*Mining Journal* (Weekly Edition), June 12, 1886.
[2]*Ibid.*, April 18, 1891.
[3]*Ibid.*, March 15, 1890.

city's sewage just prior to 1890 and in March of that year, after many meetings between city officials and the Deer Lake Company, a price of $30,000 was arrived at for the city to purchase the Deer Lake Furnace dam.[4] The purpose of this transaction was to lower the river level through the elimination of the dam, thus enabling the sewage to be run off into Deer Lake by way of the Carp River. The dam was eight feet high, and had provided the Deer Lake Company with a valuable water power which had saved $10,000 in annual operating costs.

The Deer Lake Company installed more boilers and steam engines to replace the water powered turbine wheels. In August, 1890, four feet of planking were removed from the dam to begin lowering the reservoir and by early September it was drained.

One stack of the Deer Lake Furnace was put into blast again, with the machinery being run entirely by steam power, and from April of 1891 to December, up to 30 tons were cast daily. But this old, inefficient stack was unable to turn a profit for the company and in late December, 1891, even though the iron run from it was ". . . renowned for its excellent product . . . ,"[5] it was blown out, ending a 23-year period of almost steady operation.[6]

Although the machinery and buildings remained intact for four more years, no move was made to make iron again and in September of 1895, all of the furnace and sawmill machinery and heavy castings were removed and sold to a Detroit scrap iron dealer. The buildings of the furnace remained standing for years.

REBUILDING THE CARP, EXCELSIOR AND NORTHERN FURNACES

Carp River Furnace

Noah W. Gray and Charles Schaffer became prime movers in the reconstruction of the Carp River Furnace when they acquired one-half interest in the plant. In September, 1889, the company had a dozen men removing the machinery for rebuilding in the shops. To help carry out this ambitious project and to help the local economy, the residents of Marquette pledged to raise $30,000.[7] Although only $12,000 was raised, the work to get the furnace back in blast went on. By February of 1890, all of the burned wooden buildings had been replaced by stone structures, and a double elevator to carry the flux, ore and charcoal to the stack top had been erected.

---

[4]*Iron Ore*, Ishpeming, March 15, 1890.
[5]*Ibid.*, April 4, 1891.
[6]*Ibid.*, September 3, 1892.
[7]*Ibid.*, August 24, 1889.

Carp River Furnace, c. 1895.

A stock shed was built which held 10,000 tons of ore and a coal shed measuring 50x150 feet was also erected. Schaffer had the charcoal contract and before the furnace went back into blast, he had received a good supply from Chocolay farmers. For fire protection, filling the steam boilers and many other needs about the furnace, a new water supply system was put in which supplied water from the Carp River by means of a pair of Smith Vaile pumps.

In anticipation of beginning casting that winter, the sand floor of the casting house was thawed and dried with row upon row of burning charcoal while ". . . the huge stack, ready charged for the fire, loomed grimly through its cavernous arches."[8] The stack fires were lit around March 1, 1891.

A few months before the Carp returned to blast, the doors of the nearby Marquette Branch Prison were opened to receive its first residents. This institution was built on property donated by the city of Marquette. One of the reasons this site was picked over others in the county was the fine growth of hardwoods covering the land. Its value for making charcoal was touted by the city and in December of 1890 the state contracted with Schaffer for the removal of 1,000 cords and to clean up acreage for farming. Schaffer, in turn, contracted with the Freeman Bros. to team the wood to the Carp River kilns for charring.

[8]*Mining Journal* (Weekly Edition), March 1, 1890.

The furnace blew out about 1891 and not until 1899 was there any attempt to put it back in blast. Schaffer and Gray leased the property in June of that year and in October it was making iron. To make the grade of iron he had made at the Excelsior plant, Schaffer found it necessary — because of previous sales commitments by the mines on the Marquette Iron Range — to obtain ore from the Gogebic Range.

The Carp Furnace made many tons of iron in the following years, and during the period from 1899 to 1905, the furnace operated almost continuously, producing up to 70 tons per day. The Pioneer Iron Company, a subsidiary of The Cleveland-Cliffs Iron Company, purchased the furnace in December of 1899, retaining Gray as manager and giving Schaffer the charcoal contract. Cliffs gained ownership in 1905, but the iron market became depressed and on December 7, 1907, the furnace blew out its final blast. The furnace and buildings were idle many years and, finally, in October, 1916, the Lake Shore Engine Works purchased the deteriorating plant for $1,500 and soon dismantled it.

Excelsior

The Excelsior Furnace, which had been idle since 1883, was open to summer rains and the winter snowfalls, and consequently the machinery and buildings deteriorated to a state of almost total uselessness. Vandals stripped the valuable engines and pumps of their gauges and everything that could be worked free, breaking what remained.[9]

The Excelsior Iron Works was sold at auction on October 31, 1889, to Hiram A. Burt and Charles Schaffer for $4,300, with the latter acquiring a one-third interest.[10] After an inspection by Burt and a foundryman by the name of Thomas Gorman, repair work was started in December. By August of 1890, it was producing iron, but in less than a month a charcoal shortage forced it to close down. Another blast was started in November when "several tons of pig iron were made," but it, too, ended prematurely. The owners' intention was to keep the furnace out of blast until the following spring.

For two more years the furnace sat idle which allowed time for a complete overhauling. By then, Ishpeming residents were anxious to hear the steam whistle of the furnace once again. Peter White and Charles Schaffer were then the owners of the plant, and on April 27, 1892, under the direction of Case Downing, its fires were once more started.[11] The plant operated well for the short period it was in blast; the 30 men employed were able to produce 30 tons per day. It shut down in August, 1892.

Although the pumping out of the 153-acre Lake Angeline in 1892 permitted the underground mining of iron ores under the lake bed, and caused the Excelsior to obtain water for operations elsewhere, it was not the

<hr>

[9]*Iron Agitator*, October 2, 1886.
[10]*Mining Journal* (Weekly Edition), November 21, 1891.
[11]*Ibid.*, April 30, 1892

reason the furnace went out of blast. It appears that Schaffer's colliers and wood cutters had done their work well and produced a large supply of fuel. He wanted to use the fuel in the making of iron before fire or decay ruined it, and it was his plan to run the furnace to use up the supply.

More years of idleness followed at the iron works and ownership again passed into the hands of others but Schaffer retained his interest in the company. His new partners were W. H. Nelson and Fred Smith of Fond du Lac, Wisconsin, who were connected with a blast furnace there, and E. A. Hyde of Chicago.

In May of 1894, Schaffer traveled the C&NW to Larenta and other kiln locations, and the DSS&A to Onota, to arrange for a large supply of charcoal. A strike at the Pullman Car shops in Chicago hindered the start of iron making, as a sufficient number of railroad cars were not available to transport the ore and charcoal. But in July, 1892, when the end of the strike was in sight, the fires were relit and the slow drying process of a new lining was started. The stack was charged and on July 14 the first iron was made.

To end the old practice of handling the fuel twice at the furnace, a large shed had been built that held up to five railroad cars loaded with charcoal. As the fuel was needed it was taken directly from the cars. A tram track was laid from the casting house to the nearby rail line, thus speeding the removal of the iron from the beds. As the pigs were removed they were graded and marked for future sales. The Cleveland-Cliffs Iron Company had a contract with the furnace owners to supply all the ore and to purchase all the iron that was made, and it was Patrick Carroll's job to oversee this arrangement.

During this blast the daily tonnage increased substantially, reaching a peak on April 18, 1895, with 69 tons, and it was not unusual for 68 tons to be cast during a 24-hour period. For the month of April, 1,852 tons were cast, with an avearge of 62 tons per day. The contract with Cleveland-Cliffs expired on July 15, 1895, but two weeks before this date was reached, the furnace was blown out, as 17,000 tons of pig iron were stocked in the yards. The iron market had declined and another slowdown had begun.

Immediately after blowing out, a crew was put to work repairing the old machinery, installing a new and much larger cast iron hot blast, and positioning a new hot blast engine three times larger than the old one. Twenty men were working on the repairs in October, 1895, and most of the stockpiled iron had been sold. On November 18, Manager Nelson put the Excelsior into blast again, making it the only furnace operating in the Upper Peninsula. About the same time electricity was introduced to the iron works, with power supplied by the Negaunee & Ishpeming Street Railway & Lighting Company. Both arc and incandescent lights were used in the buildings.

With the furnace producing over 70 tons of metal per day, a shortage of fuel developed in January, 1896, and the blast was stopped for over a month. The kiln stations had used all of the wood on hand to make charcoal and the bad roads in the woods, caused by a shortage of snow that winter, prevented the hauling of more cordwood to the kilns.

Shortly after going into blast in February, the furnace "explosion doors" flew open and fire rolled out into the roof rafters, giving the appearance of certain doom for the buildings. The furnace whistles screamed a warning, as did those of the nearby Lake Angeline Mine, but when the firemen arrived the furnace hands had the flames under control. The first opening of the explosion doors, which was considered normal under certain conditions, was always a good scare for the crew, but it did not happen often and confusion and near panic was the result for a while when they did drop out.[12]

The Lake Superior Iron Company had the contract with the Excelsior Furnace in 1896. In October, with 20,000 tons stocked in the yards on every available foot of ground and production reaching 75 tons per day, the furnace was blown out. An order was received from Cleveland that month for 5,000 tons of iron, but this still left an enormous surplus, and it was uncertain when and if the furnace would return to blast.

Northern Furnace — Harvey

Frank B. Spear and a group of Marquette capitalists including Nathan M. Kaufman, James M. Wilkinson, John M. Longyear and J. C. Reynolds, gained control of the Northern Furnace at Harvey in 1890[13]. With a capital stock of $100,000, refitting was started at the furnace where fires had been out since 1867.[14] Under the name of the "Northern Furnace Co.," it was again changed to produce charcoal iron. A new blowing engine was installed in expectation of a long run.

On January 15, 1891, it was put into blast and produced from 41 to 72 tons of iron per day. But a poor iron market and a waning supply of charcoal limited the furnace only a 10-month run. W. L. Wetmore had started a small charcoal industry in Munising in the fall of 1890, having a sawmill and five kilns in operation, and relied almost solely on the Northern's demands for a market. The Duluth, South Shore & Atlantic Railroad provided transportation for his charcoal to Harvey. His plans were to add more kilns to the battery but whether or not this was carried out is unknown. Between $3,000 and $5,000 were paid each month by the furnace for labor, charcoal made by local farmers and the charcoal industries in Alger County; the unexpected shutdown in November of 1891 hurt many people financially.

[12]*Mining Journal* (Weekly Edition), February 8, 1896.

[13]*Iron Ore*, April 19, 1890.

[14]*Mining Journal*, March 16, 1878.

It was said that an "immense" amount of pig iron was made during the run but it actually was 12,500 tons, still an impressive figure.[15] After the furnace was shut down, most of the iron remained piled in the yards as the iron market was sated for the time being. The Northern once again went into many years of idleness.

The Lake Superior Chemical and Iron Company with furnaces at Gladstone, Boyne City, Ashland, Wisconsin; Manistique and Newberry, took an interest in the Northern Furnace in 1907 and on January 4, 1910, the "wind" was turned on and iron making was started.[16] Fifty new kilns had been built by the company and filled with cordwood, but financial troubles soon developed and the furnace was only in blast a short time. The wood in the kilns was later shipped to the company's Newberry furnace. Although there were plans to relight the fires in 1912, nothing ever came of them, and the plant was stripped of all machinery.

WESTON FURNACE COMPANY

Abijah Weston and other well known lumbermen, who owned vast tracts of hardwood timber lands in Schoolcraft County, joined with the owners of the Pine Lake Furnace Company of Ironton, Michigan, in February of 1890. They planned to build a large charcoal iron furnace to make use of the available fuel. The site selected was near Lake Michigan and harbor in Manistique.

Articles of incorporation were drawn up on April 24, 1890, and the contract for erecting the furnace in Manistique was let to R. H. Cherrie. The stack of this well planned furnace was built to have a daily capacity of 100 tons, and on April 30, 1891, it was blown in.[17] The following month the stack was averaging 75 tons of charcoal pig iron daily, and it was expected that production would double.

With a view of using ore from the Marquette Iron Range, the company projected a railroad to be built from Manistique to Negaunee to save an estimated 35 cents per ton in shipping the ore and to open a larger area for securing charcoal. Articles of incorporation were filed for the Manistique & Northwestern Railroad in May of 1891, with a capitalization of $1,600,000, and in the fall and winter of 1891-92, the preliminary survey was made. Plans were also made for an ore dock to be built at Manistique as Abijah Weston already owned a fleet of lake carriers. By April of 1892, right-of-way for three quarters of the proposed route had been secured, but some resistance was met in Negaunee where the track was to run through a thickly populated part of the city to reach Teal Lake. Here, a station was to be built and a branch track run west to the nearby Cambria and Lillie mines where the railroad would terminate.

The Manistique Furnace Company contracted with the Davis Mine in

---

[15]*Mining Journal*, (Weekly Edition), January 2, 1892.

[16]*Daily Mining Journal*, January 6, 1910.

[17]*Mining Journal* (Weekly Edition), March 7, 1891.

Views of the furnace at Manistique, published in the *Manistique Record*, 1904.

Negaunee for iron ore in 1894, and went into blast for a short time, blowing out before the close of the year.[18] It remained idle until the summer of 1897, when it went into blast with ore mainly from the Negaunee mines. On one day in mid-July of that year, 100 rail carloads were shipped to Manistique, coming from the Queen, Lucy, Lillie and Cambria mines. On the days following, until that fall, 40 carloads per day left Negaunee for the furnace. The furnace company had also contracted for 10,000 tons of ore from the Richmond Mine near Palmer on the Cascade Range.

The plans for the projected railroad lay dormant for some time but it was reported by George A. Newett, Commissioner of Mineral Statistics, that 20 miles of the road were completed by 1896; the rolling stock had not yet been secured.[19]

The starting point of the railroad was in South Manistique and it ran north ending at Ackers. An extension was completed to Shingleton on December 31, 1898, where it made connection with the Duluth, South Shore & Atlantic Railroad. However, it was used more extensively in the white pine logging industry than for iron ore and charcoal.

The chemical plant of the Newberry Furnace was dismantled in 1897 and arrangements were made to have the entire plant moved to Manistique to make wood alcohol at the furnace there.

Joseph H. Berry, who was one of the owners of the Burrell Chemical Company, which operated the furnace's charcoal kilns in 1899 (and part

---

[18]There was also a Davis mine on the Gogebic Iron Range.

[19]Newett, George A., *Mines and Minerals Statistics — State of Michigan*, (Lansing, 1896), p. 143.

owner of Berry Brothers, a Detroit varnish making concern), purchased the Manistique Furnace in the spring of that year. A few months later, in February of 1900, the Griffith Car Wheel Company had charge of the furnace and hired W. H. Nelson of the defunct Excelsior Iron Works to manage it. He put it in blast in May.

The Manistique Furnace had times of prosperity, but mostly it went through the same fluctuations as the other furnaces in the Upper Peninsula, with shortages of charcoal and over-stocked yards of pig iron. Consequently, it passed from one owner to another. The Manistique Iron Company owned the iron works in 1904 and at the time 500 men were employed and 100 tons of iron were being made daily. In 1913, the plant was being run by the Lake Superior Iron & Chemical Company, and later, in 1922, it was managed by the Charcoal Iron Company of America, and was still in blast.

THE IRON CLIFFS PIONEER FURNACE SOLD

The controlling interest in the Iron Cliffs Company was purchased by the Cleveland Iron Mining Company in February, 1890, and in 1891 The Cleveland-Cliffs Iron Company was formed through a merger of the Cleveland and Iron Cliffs companies. Cliffs gained control of the Salisbury and Barnum Mines in Ishpeming, the Foster Mine a few miles southeast, 53,000 acres of valuable timber and mineral lands in the area, and the working Pioneer Furnace in Negaunee. Under the new management, business at the furnace carried on as usual.

The Pioneer's coal shed burst into flames in May of 1891, and when the walls weakened and split open, charcoal ran out in a steady stream, covering tiers of pig iron nearby. Sand was shoveled on the burning coal, completely covering it, but the charcoal continued to smolder underneath the sand for many days. When it was finally extinguished, it was found that the burning charcoal had melted the ends of many tons of pigs together. The damage done to the shed and iron was extensive and 250,000 bushels of charcoal were lost, but the furnace remained in blast and made iron until mid-1892.

On June 11, 1892, the breakdown of one of the blowing engines stopped No. 1 permanently, and later that month it was reported that, "The old furnace is now completely worn out and unfit for further service . . . ."[20] Then, on July 27, 1892, No. 2 was blown out, apparently for the last time, leaving piles of ore still stocked at the iron works.

In December, Patrick H. Carroll and Samuel Redfern, residents of Negaunee and both knowledgeable in the field of making pig iron, leased the Pioneer Furnace to run it for the last time. With Case Downing as founder, the first cast of iron under the new management was run out of

---

[20]*Mining Journal* (Weekly Edition), June 18, 1892.

No. 2 stack on January 4, 1893. These men operated the furnace with the understanding that they would repair all damages to the plant which occurred while it was in their hands. Fuel was drawn from the Houston Kilns, located between Negaunee and Palmer, where 1,100 cords of wood were already banked, and from kilns at Ford River, Goose Lake, and Section 6.

The furnace produced a fair quantity of iron but at a reduced rate due to a constant fuel shortage, and in May, 1893, the condition of the machinery and stack was such that Carroll and Redfern feared a breakdown could occur at any time.[21] Not caring to expend any money on expensive repairs, they used up the stock of ore and charcoal on hand. On May 18, 1893, life ended for the oldest iron works in the Upper Peninsula.

The death of Andrew Buckely in Negaunee in November of 1898 revived memories of the early iron making days at the Pioneer, where this huge German-born furnace hand would perform great feats of almost superhuman strength. It was Andrew Buckely who wheeled out of the casting house of the No. 1 stack, the first pigs made in the late 1850s, and for years after he worked at this plant, doing the heavy work which was always encountered at a place of this nature. Flat-bedded wheelbarrows were used to move the pigs to the stock yard or loading dock, and many times during his long employment at the furnace he had wheeled loads weighing one and a quarter tons. Many times two men would try together unsuccessfully to move the same load. The ''furnace giant,'' as he was called, was born in 1823 and had also worked at Collinsville and at Morgan.

For over two years the iron works sat without turning a wheel. Then, in October of 1895, most of the heavy machinery was sold to a scrap iron dealer in Detroit. The boilers and the cast iron ovens (which had to be broken up by blasting) were also sold for scrap, but some of the better machinery was salvaged for future use at Gladstone. Most of the buildings were left intact. For a while, Louis Corbett stored coal there and in one of them a bonemeal mill was put into operation.

PUFFER

The Cleveland-Cliffs Iron Company began work on its new furnace at Gladstone in April, 1895. This furnace was the last blast furnace built in the Upper Peninsula during the 50-year period from 1848 to 1898, and, as the company's furnace at Negaunee had ceased operations, its name, the *Pioneer*, was transferred to Gladstone along with *Puffer*, a little switch engine that had seen many years of service in the Pioneer yards in Negaunee.

The furnace site was located one and one-half miles north of Gladstone on a small island hugging the shore of Lake Michigan, and considerable dredging was done to deepen a channel for the big boats that were

[21]*Mining Journal* (Weekly Edition), May 17, 1893, and *Daily Mining Journal*, May 20, 1893.

expected. The sand and gravel from the channel provided fill to make a railroad grade to the mainland and also provided the means to build up the level of the island which was quite low, bringing the furnace level up to nine feet above the lake. In all, 50,000 cubic yards of material were scooped out of the channel.

To supply charcoal for this modern stack — which stood 60 feet high and had boshes of 12 feet — 40 kilns of 62-cord capacity, each measuring 31 feet in diameter, were built at the iron works in a double row. These eventually supplied 65 percent of the fuel and, like the old Northern Furnace at Harvey which was partly supplied by neighboring farmers, the Gladstone Furnace received the balance of its fuel in the same manner.[22] The Cleveland-Cliffs Iron Company owned 8,500 acres of hardwood lands 30 miles from Gladstone with rail access, and 12,000 acres were set aside for fuel at other locations.

The newly designed hearth was lined with one-inch steel plating while cooled with an external spray of water to extend its life and the furnace was fitted with five tuyeres. The two blowing engines from the Pioneer at Negaunee worked the hot blast, and air for the blast was heated in two Cowper stoves which required over 130,000 bricks each in their construction.

Water sprayers were also placed on each side of the 50-foot by 90-foot casting house to moisten the sand in the pig bed. This was a big improvement over the older method which used hoses. A dock which measured 150 feet wide by 600 feet long fronting in water 17 feet deep was built near the casting house on the channel.

The Gladstone Pioneer was completed in early February, 1896, but the chemical plant connected with the furnace was not finished and the start-up was thus delayed for a short time. The company also erected 10 two-family houses for the furnace hands and a home for Patrick Carroll, the favored founder, near the iron works.[23] A more dedicated founder was not to be found in the Peninsula and wherever he worked furnace efficiency improved. Founders like Carroll were well liked by the community and were considered the backbone of the charcoal iron industry. In Gladstone, the company also built homes for Austin Farrell, the manager; C. V. R. Townsend, clerk, and for George Slining, master mechanic.

Construction of the furnace allowed an easy switch to coke if the owners so desired, but when first put into operation, charcoal iron was made. The kilns were fired up on March 10 and the smoke from each was piped to

---

[22]*Iron Ore*, March 21, 1896.

[23]"Patrick Carroll died in 1898 at the age of 53 in Gladstone after working 33 years in the charcoal iron industry on the Upper Peninsula as a founder. He started at the Negaunee Pioneer in 1865, which was then owned by the Iron Cliffs Co., one of the parent companies of The Cleveland-Cliffs Iron Co., and from there moved to the Deer Lake furnace for two years. This was followed by employment at the Morgan Furnace, then one year at the Escanaba in 1874. After a brief try at mining in the Lake Superior Iron Co.'s mine, where he was injured, Carroll returned to the Deer Lake and worked it until it closed. He then worked at the Negaunee Pioneer again until it closed, then went to the Excelsior for one year. From there he moved to the Gladstone Pioneer." — *Mining Journal* (Weekly Edition), January 15, 1898.

Pioneer Furnace, Gladstone.

the chemical plant which stood several hundred feet from the furnace, being drawn by a vacuum fan into three batteries of wooden tanks (five tanks in each battery). Inside the tanks, water-cooled copper tubing condensed the wood alcohol from the gaseous hardwood smoke.

Mrs. Townsend, wife of the clerk, lit the initial fires in the furnace on April 16, 1896. Shortly after, one of the ponderous blowing engines was put in motion and the first blast was started that put the furnace in a race to capture as much of the limited market then available for charcoal iron.

The furnace smelted Foster Mine ore for a time starting in June of 1897 and later, in February of 1900, it was producing as high as 130 tons of charcoal iron per day, averaging 120 tons daily. The chemical plant and the furnace were consuming a total of 240 cords of hardwood and manufactured charcoal each day, and was running off wood alcohol at the rate of 14,000 gallons a month.

The Gladstone Pioneer was not without its destructive fires. Lightning struck the No. 1 stack's chemical plant in 1904 and when the fire was finally put out, the building was completely destroyed. Though the furnace was saved in the $150,000 fire, the company carried no fire insurance on the chemical plant, and because of it the furnace remained out of blast for some time.[24]

---

[24]According to Harland Hatcher's *A Century of Iron and Men* (New York, 1950), "It continued in operation until 1922 when the supply of wood in region was exhausted. In 1933 . . . it was pulled down and sold for scrap." (pp. 203-204).

A "slip" in the Pioneer Furnace, Gladstone, in January of 1898 — a common occurrence at all iron works. From *Lake Superior Mining Institute Proceedings*, 1903.

THE WATER COMMISSIONERS ACT

Because of previous experience with vandalism while the plant was idle, Manager Nelson of the Excelsior retained a few men when it shut down in 1896, and under their care it was kept in good repair and ready to make iron on short notice. New orders for iron caused the furnace to return to blast in August, 1897. Except for being shut down for one week in January of 1898 because of a broken elevator, and short interruptions because of gas explosions in the stack (which started buildings burning and caused fires in the nearby woods), the furnace remained in blast until that fall.

Enough orders for iron were received regularly to keep the old stack in blast, leaving very little iron stockpiled in the yards. The problem of a water supply for the furnace arose again and, under the watchful eye of the Ishpeming water commissioners, on November 4, 1898, the plant was shut down.[25] The commissioners considered the furnace a dangerous drain on Lake Sally, the source of the city's own water supply, and although they regretted shutting the plant down and putting the furnace hands out of work, they felt they had no alternative. The owners of the furnace petitioned the commissioners in December of 1899 to supply three one-inch water lines to operate the furnace and again were turned down. The inability of the furnace owners to secure a proper water supply thus ended the last attempt to make pig iron in the city of Ishpeming.

"The old charcoal furnaces whose remains now survive only to indicate the primitiveness of the industry and the energy of its promoters . . . ,"

---

[25]*Mining Journal* (Weekly Edition), November 12, 1898.

is as true today as when this statement was made in 1902.[26] When viewing these long deserted communities with their rows of "Michigan" dugout basements, one wonders how these people accomplished what they did during the long, bitter cold winters, and the summers of swarming mosquitos and black flies. A good example of a location which is now devoid of life is Morgan where only two of the many basements had masonry walls which, with wells and other excavations, are still visible. Furnace builders apparently went over their lands in search of the right terrain and a stream that would fit the needs of a blast furnace. Consequently, some of the stacks were erected in rather remote locations.

The Fayette Furnace complex on the Garden Peninsula in Delta County is being rebuilt by the State of Michigan on an ongoing basis, bringing it closer to its original state each year. A visit to this state park is a "must" for any person interested in the iron making history of the Upper Peninsula. The remains of very few of the other blast furnaces are to be seen today, though the Champion, without any upkeep, is probably in better shape than any as it is fenced in by its present owners. About 30 feet of the Morgan stack is still standing with a roofless room attached. Each year a little more of this relic is destroyed by people climbing on it. The Greenwood, being quite difficult to locate, is a large pile of rubble with broken cast iron pipes reaching out, while other furnaces stand in various stages of decomposition, returning to nature.

Countless charcoal kilns can be found scattered throughout the Peninsula — most of them merely circular heaps of native rock, brick, stone, or whatever was used to construct them. Onota, east of Deerton, probably has one of the better undisturbed batteries of fallen kilns and stone buildings to be observed in the area. Some of the iron kiln doors are still buried there.

★ ★ ★ ★ ★

The volume of fuel needed to reduce the ore to metal was probably the biggest single factor in there never being a successful charcoal or, for that matter, coke, soft coal, or hard coal furnace to profitably survive in the Peninsula. The ability to obtain ore and flux was no problem but shortages of fuel in any form stopped many blasts prematurely, a disheartening experience for any furnaceman.

It has been estimated that in making the 2,000,000 tons of pig iron in the Peninsula prior to 1902, 166,666 acres of timberlands were cleared for the 5,000,000 cords of wood converted to charcoal in the several hundred kilns. In 1867, it was figured that 2¼ cords of wood produced 100 bushels of charcoal, using maple and birch. With this in mind, it appears that approximately 222 million bushels of charcoal were used in the iron making process.

---

[24]*Proceedings of Lake Superior Mining Institute* (Ishpeming, Mich.), IX, p. 75.

Two hundred and eighty-nine furnaces were in blast in the United States in 1899. Of these, 191 were fueled by bituminous coal and coke, 68 were fueled by anthracite or a mixture of anthracite and coke, and only 30 were using charcoal. The last charcoal iron furnace to operate in the United States was the Newberry Furnace, which went out of blast in September of 1945.

CHARCOAL ANECDOTES

The following anecdotes by Thomas Clancey of Ishpeming originally appeared in the July-October 1921 issue of *Michigan History* quarterly. They are quoted at length.

"It is not to be presumed that this rather jumbled collection of tales and incidents will be an adequate treatise of the wit and humor of the days when the charcoal furnaces flourished in Marquette County. Far from it. Neither will it be a psychological analysis of the motives that prompted the execution of this section. It will be the humble endeavor of this remote chronicler to set down here a few incidents and stories in order to help us gain an impression of the character of the entertainment these men had to indulge in to tide themselves over the periods of work and play. The work was hard and exhausting, the recreation and amusement had to originate in the active brains of the furnace community. Traveling theatricals never visited them, and when navigation closed in the fall they were left to their devices for diversion during the winter.

"The very instability of the science of making pig iron and the crudeness of the plants were a factor in increasing the cost of production. The sustaining factor, of course, was the high price of manufactured pig iron.

"Naturally, an industry of this character which carried such uncertainty and (required such) laborious tasks, attracted a venturesome class of labor, trained in the craft of iron making. They followed furnace work and went from one place to another as fortune and opportunity warranted. They traveled about a great deal, being attached to foremen and 'bosses.' When their leaders were given charge of new furnaces or were sent to solve a problem of getting an old furnace in running order, the leaders took with them the best workmen to aid them in various departments. In this was a spirit of loyalty and pride in the work developed which became in many instances an opportunity for rivalry, sometimes reaching the point of a fight before the supremacy of the gang was really established. However, these men took real pride in the furnace where they were employed, and, due to smallness of numbers, and their ability to draw around them kindred spirits, reached a point of efficiency much sought and seldom reached by our modern industrial establishments.

"The men who were employed in the charcoal furnaces in this county were in the main New York state men from the Lake Champlain district,

Charcoal kilns near the mouth of the Carp River at Marquette as they appeared in the 1920s.

where they learned the iron-making business — Irish, Scotch, Germans, and French Canadians. The actual furnace work was done by the native Americans and Irish, while the French were the choppers and prepared the charcoal. When the kilns were located at any distance from the furnaces, the settlements at the kilns were almost entirely peopled by the French. This work was done by contract, and although the contractors had little capital, credit at the company office and store supplied this want. Communities grew up about the furnaces and each developed its own social entertainment. The entire male portion of the family worked at the furnace and the work and fun was bound together. The son followed the father and learned the trade from him. Furnace work, like mining, ran in families. At one time in this county, five Dundon brothers operated five furnaces.[27]

"The old Pioneer Furnace at Negaunee numbered a choice collection of rough and ready humorists in its employ. The late John Downing, of Marquette, was in charge of this furnace and enjoyed participating in a joke even during working hours. A great 'stunt' around a blast furnace was the accurate use of a water pail. For fire protection, buckets of water were placed around the plants and on the bridge near the stack. Pails of whitewash were often left around, as whitewash was the usual decorative scheme around furnaces. An accurate bucket-thrower with one arm could land the water at a given point with neatness and dis-

[27]Richard was the founder at Clarksburg, Patrick at Collinsville, James at Champion, Maurice at Deer Lake, and Lawrence at Morgan.

patch. This was a form of amusement used on the occasion when pompous individuals were being shown around the furnace. An itinerant preacher who came to Negaunee and railed lustily at all the human frailties of the citizens, was being taken through the furnace, and while he was admiring the novelty of the casting-house, a bucket of whitewash thrown from an unseen corner smeared his clerical habiliments from head to foot.

"On another occasion the superintendent was escorting several flashily dressed Hebrew gentlemen, wearing silk hats, through the works, and simultaneously two buckets of water drenched them both.

"The Deer Lake Furnace, which was built in 1867 and made its first cast of pig iron in 1868, was run on a rather peculiar method for a business proposition. In 1868 and 1869, the owners of it, Messrs. Hungerford and Ward, being of pronounced religious convictions — Mr. Hungerford being a deacon of long standing — objected to the performance of any work about the plant on Sunday. When ore is heated in a furnace, one of the prime factors of a successful cast is heating up the stack and keeping it at the required heat continuously. Due to the religious manner in which this furnace was operated, the fires were banked at midnight Saturday and nothing done until Monday morning. During the interim the stack would chill and would take a day or two to heat up again. Patrick Carroll was the founder at Deer Lake and was one of the most skilled in this region. James Clancey had charge of the fires and blowing engine. One Saturday night the furnace was running at a low degree of heat, due to the large amount of rock in the stack, and in danger of chilling and "hanging up," thus injuring the furnace for some time. Carroll and Clancey decided to increase the heat and put through a blast regardless of orders. To allow a "hang up" to occur was not only bad business policy, but also a reflection on the ability of the furnacemen. They put through the cast on Sunday, saving the plant a long shut-down. When the proprietors returned from Marquette on Monday and discovered that their orders had been disobeyed, they gave a lecture to the two offenders on their irreligious conduct in saving the property. Carroll and Clancey sent back a heated blast, drew their time, and left to pursue their trade where religion was not carried to such extremes and their reputation as foundrymen not impaired thereby.

"The furnace employees were not in the van of the present prohibition movement. Liquor was used rather freely, and it was a poor wedding or christening where beverages were not provided. A birth in the family was a signal for the advent of a keg of beer in the location. At the Deer Lake location the ideas of the management in regard to liquor were well known. A certain Mrs. Harrington gave birth to a male heir, and her husband, wishing to express his joy but still having in mind his employers' views on the matter of liquor, advised the Ishpeming storekeeper to put a keg of beer in a flour barrel and send it with the store team. Joshua Hodgkins, father of Gilbert Hodgkins of Marquette, was the company detective, and he ferreted out the keg and prevented the entrance of it into the Harrington house — much to the discomfiture of Harrington and his friends.

"In 1871, the operators of the Morgan Furnace conceived the idea that malleable iron could be hammered out into blooms from the ore as it ran from the stack. This was a brand-new idea to charcoal furnacemen and the success of it was much doubted. However, the operators of the furnace, in order to try out the plan, imported several Swedish iron men and brought them to the plant. Under their supervision, the plant was remodeled, rotating furnaces were installed for the grinding of charcoal to fine powder to generate the necessary degree of heat, and a large amount of money was expended. The Swedish iron makers were very secretive as to the method which they were about to employ and jealously guarded all their movements so that the local furnacemen would not be able to steal or copy the system they used. Several months were consumed in the preparatory work, the equipment was put in running order, furnaces were heated, and the ore placed therein, when the valiant Swedish iron makers carefully drew their money from the office and vanished into the neighboring woods, never to be seen or heard of again. Nothing daunted by this failure, the management brought Mr. Jones from Pittsburgh, the so-called expert in the manufacture of malleable iron by this method, to the Morgan Furnace. He brought his wife and they took up their residence there. He inspected the plant, and said that malleable iron could be made by this method but the machinery was not installed properly. The changes he required were made, consuming another three or four months. In the meantime, his wife had returned to Pittsburgh, and when the furnace was ready to operate and the ore was heated and all the construction had been carried out, he suddenly received a telegram announcing the serious illness of his wife. He left without divulging the great secret which he carried in his brain.

"Still persevering, Messrs. Ely and Donkersley secured the services of a number of iron puddlers at the Rolling Mill Furnace, then operating at Marquette. After the arrival of the puddlers, who also brought several kegs of beer to sustain them during this scientific ordeal, the furnace was heated, the iron placed therein, and the rotation of the furnace commenced. As the work of melting the ore progressed, the Morgan furnacemen noticed that from time to time the Marquette operators slipped into the furnace large quantities of iron chips and shavings. Naturally, as the furnace was rotated, the chips collected the molten metal and it began to take the appearance which was desired, of the large round ball. The beer had been unstintingly passed out to the impoted scientific investigators, and the appearance of the molten ball was greeted with loud cheers and the feeling that now the success of the method was established. When the molten ball was carried over under the steam hammer, which would pound it into the much sought for blooms, at the first drop of the hammer the molten mass and the chips and shavings scattered like the autumn leaves about the furnace, and the great secret eluded its pursuers.

Excelsior Furnace, Ishpeming, in blast with a yard of pig iron.

"Social activities of a small character were quite common during the winter months. At the completion of the Morgan Furnace casting house, it was thoroughly whitewashed and boarded up. Evans' orchestra from Marquette was imported, and a dance and supper were furnished at $5.00 a couple. They were a sociable lot about the furnace location, and thought nothing of loading a sled full of people and driving from Morgan to Champion to attend a dance. A great rivalry existed between the different furnaces. The organization that produced the largest tonnage of pig iron considered itself the champion of the district. Decorated brooms were put up and contested for. At the end of the cast, a dinner would be held for the contesting teams, liquid potations would be indulged in, and the winning team would ofttimes be compelled to maintain its supremacy by the appeal to bare knuckles.

"The furnace community life, although it contained little of heroic or startling nature, still had an air of quaintness and partook in a large measure of the real pioneer development of this county. If we can catch a glimpse of the daily life of the people who engaged in this industry, we have done about all that could be expected in this remote year of 1921."[28]

---

[28]*Transactions of the Lake Superior Mining Institute* (Ishpeming, Mich., 1926), XXV, pp. 252-258.

Chapter IX

ACCIDENTS IN THE INDUSTRY

The charcoal iron industry like the mining industry was not without its share of accidents and human tragedies. Following, chronologically, are the worst recorded accidents of Upper Peninsula blast furnaces.

★ ★ ★

The loaded water balance at the Negaunee Pioneer Furnace took the first life, that of Edward Conners, in May, 1869. Conners was cleaning machinery at the bottom of the water balances and got his head caught between the balance and the frame, where he was crushed and killed instantly. It was said that Conners was a sober and industrious person who had saved nearly $1,500, and was about to embark on a journey to the West in a matter of days to purchase land for a farm where he was to live with his wife and two children.

★ ★ ★

The Pioneer was again the site of an accident in August, 1873 — probably the worst to happen at any of the furnaces. On an elevated track, two loaded ore cars were being pushed to the crusher, but part way there the track sagged and the old, weakened timbers groaned. Unable to move the cars further, 14 men set to work bracing up the slowly cracking timbers. Suddenly, the trestle gave way under the heavy load and the cars fell, spilling the chunks of ore onto the men.

Eleven of the men were caught in the rain of ore, cars and timbers, with three — James Curley, Maurice Merrill, and a man named Lemere — killed instantly. A Mr. Weggens and James Murphy suffered broken legs. Peter Lamere lost his scalp, and the others — Maurice Valnette, Pat Kane, Michael Curley and Pat Katen — were all seriously injured.

★ ★ ★

Providence was with a man at Fayette who was charging one of the stacks in March, 1874. While dumping a load of coal he fell into the hot smoking top, but was rescued unharmed except for a slightly burned arm and a thorough scare.

★ ★ ★

At Forestville, in May, 1876, J. L'Huillier, believed to be the L'Huillier that made iron on contract at the plant in 1874, drowned in Dead River while cutting a log above the dam in the process of cleaning out the reservoir.

★ ★ ★

While dismantling the Escanaba Furnace in July, 1879, George Reed was killed taking down the brick walls of the stack for Atkinson & Stevens.

Fifteen-year-old William Smith, employed at the Pioneer in June, 1880, as air hoist operator, fell a distance of 42 feet when he stepped off the elevator and missed a beam. Landing on another beam he broke a leg and arm but recovered.

The Marquette & Pacific (Rolling Mill) Furnace was the scene of a startling and painful accident that took place in June of 1881. Thirty-five year old Ole Olson was working near the furnace when a slip in the stack forced the hot blast cleanout doors to swing open. The flames rolled out, caught Olson, burning him and the clothes off his back. Men working 60 feet away had to run from the fire, but others rescued Olson and he was brought to the County House for treatment. He recovered from his frightful experience.

Normally a "slip" in the stack was of no consequence except for a period of billowing black smoke and possible scattered fires about the iron works. In August of 1882, Moses Goodrich was charging the stack at the Pioneer Furnace. He had raised the bell and was in the act of dumping the ingredients in the stack when the inner flaming mass suddenly dropped, forcing gaseous flames out the opening which completely covered the unsuspecting man. He died from the burns in less than a week.

A French Canadian named Adolf Caron was cleaning near the top of the Deer Lake Furnace in December, 1882, and when the heat and smoke became unbearable, he stepped back too far and he went over the edge and fell to the ground. Caron died in a short time and his body was returned to his home in Quebec.

John Cheavins, working the night shift at the Vulcan Furnace as hoist operator, failed to put blocking on the wheels of the rail car he was unloading. The car rolled forward and pinned him against a column. Severely injured internally, he was taken to the Nelson Hotel where Dr. Voorhies and Ferrerd examined him, but his condition was beyond their help and Cheavins died that evening.

An experimental kiln at the Vulcan Furnace, designed to use the exiting smoke in the manufacture of wood alcohol, blew up in December of 1885. Leonard Jenny, in charge of the kiln, was struck by flying debris and was killed, and the kiln was completely demolished by the explosion. At the Vulcan Furnace again, in January, 1887, John Grant was killed while shoveling ore from a pile of frozen ore, when seven tons fell on him.

★ ★ ★

Fifty-six year old James A. Root, founder and manager at the Pioneer Furnace in Negaunee, was on his way to work at the iron works on the morning of March 8, 1886, and while walking through the snow-clogged yards where the Chicago & North Western Railway was pushing cars of charcoal over the icy rails, he slipped and fell on the tracks. At that instant a bell-less engine with a string of loaded cars started backing into the yards on the same track where Root lay stunned. The first car caught him with its brake-beams and dragged him down the track, the wheels inches away from his body. Furnace hands standing nearby heard his yells for help and waved down the train, but not before the brake-beams had crushed him. Root had worked a total of 40 years in furnaces throughout the country.

On the eve of Northern's furnace startup on January 4, 1910, a sticking valve came to the attention of chief engineer Charles Dolezell who, with three other men bearing a "lighted torch and lantern," proceeded to the trouble area. A more experienced furnace hand warned of the danger of gas from a furnace just after its initial light off. The valve was repaired in a short time and the four men started down, but by a different route that would have taken them through the machine shop and engine room. As they entered an upper floor of the engine room by means of a 20-foot-long walk-way from the top of the stack, an explosion took place that knocked them flat, and sent bricks and glass showering down on the workers below. Thirty-five year old Dolezell, 28-year-old John Trudeau, and George Nowak were killed instantly. John Dasey, the fourth member of the crew, escaped but was in a serious condition for many weeks. When the furnace finally made its final cast on January 9, Dasey still could not tell what had happened.

1852 photo of Marquette Iron Company Forge.

EPILOGUE

MINERS AND MINING

Iron ore discovered in Marquette County and copper found on the Keweenaw Peninsula were the main reasons for the early development of Northern Michigan. Today, more than 130 years later, they are still important to the Upper Peninsula. It is true that silver, gold and other valuable metals have been found here, and mined to some extent. Many of these mines were worked on speculation and could not be considered to be of lasting economic benefit. Regardless of the success or failure of the charcoal iron furnaces on the Upper Peninsula, the demand for the high grade iron ore from the different ranges continued and expanded. The mines developed from pits to shafts and drifts, and then slippery ladder ways that went down into the red earth hundreds and later thousands of feet. The Mather Mine "B" Shaft in Negaunee, the last working underground operations on the Marquette Range, presently reaches straight down to a depth of 3,627 feet, while the only other underground iron ore mine in the Peninsula, the Sherwood Mine on the Menominee Range at Iron River, goes down to a working depth of 1,900 feet.

The first iron ore mines were merely open pits or "cuts" as they were sometimes called, which in time made enormous holes in the almost solid rock. The miners working these pits received $2.00 a day in 1865 for 10 hours of backbreaking work, and paid $20.00 a month to the company for board. Open pit mining consisted of blasting and breaking the ore into pieces which could be lifted by the miners into one-horse carts that hauled the ore to the railroad, in some places only a few hundred feet, then ". . . thrown into six-ton four wheel cars, (and) carried to the wharfs at Marquette, where they are unloaded into pockets or hoppers, shoots, and thence into the vessels . . . ."[1] These open workings soon reached depths of nearly 200 feet and, to stay with the rich ore, it became necessary to start tunnelling. The Cornish, with experience gained in the tin mines of their native England, adapted to this type of mining with ease. On June 25, 1870, a *Mining Journal* reporter had this say about the Jackson Mine in Negaunee:

". . . We peered down the yawning pits 180 feet deep, walked through 2,000 feet of tunnels, one of them 1,000 feet long, with soot-begrimed miners picking away, by candlelight, at the rich ore . . . . Coming out of the mine later . . . we stood on a hill near the brink of a pit, and on the opposite bank a miner, who had scrambled up was shouting to the utmost compass of his voice 'strong hole of glycerine!' . . . we learned that a 'strong hole' was one from which the ore was not likely to scatter."

---

[1] *Lake Superior Mining & Manufacturing News*, July 4, 1867.

In the first half century of iron mining on the Marquette Range the list of mines is almost endless. Each company had one or more mines in operation. There were continual starts in the mining industry, though most of the smaller companies failed. With the transition from the open pits to the underground mines (the Republic Mine went underground in 1875-76), accidents increased at an alarming rate. For example, in 1887, there were 24 recorded mining deaths in Marquette County and possibly more; surviving were many more men whose bodies were broken and mutilated.

With the many Americans, Cornish, Scandinavians, Italians, and representatives from nearly every nationality of Europe, and even a few French-Canadians lured out of the woods, the sad and unfortunate death of a fellow miner was quickly forgotten. Most were total strangers and could not even communicate in the same language. The accidental mining deaths and disabling injuries did prompt the mining companies (nearly every mining company of any pretention)[2] to build and staff mine hospitals that provided the medical attention needed by their sick and injured employees. The miners had access to comfortable wards where they were looked after by excellent nurses and tended to by the best doctors for the monthly payment of one dollar.

It was a practice in those days to have an older son of a miner work along side his father in the mine, and it was not unusual to find a very young boy working in a very responsible position. One of the deaths recorded in 1887 was that of 13-year-old Willie Roberts, tender of skip ''A'' shaft, whose father was pit boss in the Winthrop Mine. One day in January, Willie's lamp went out and, while trying to find his way to the surface in total darkness, he stumbled into the open shaft, fell a great distance, and drowned in 85 feet of water that had collected at the bottom of the shaft. In March of the same year, 16-year-old William Jones, filling the same position in the Lake Angeline Mine, fell in the inclined shaft and rolled down to the third level, a distance of 125 feet. Then, with both legs broken, each in three different places, and his body badly bruised, young Jones managaed to yell for help and was quickly transported to the hospital where his bones were set.

Thirteen other miners working in the Gogebic and Menominee Iron Range mines died from mining accidents in 1887, bringing the year's death toll to 37 or more. Multiple deaths were quite common. While blasting in the Sturgeon River Mine on the Menominee Iron Range, a stream of water was tapped that rushed into the underground workings in a torrent, drowning eight miners. They ran to escape this underground horror and in their haste to outdistance the rising water several life-saving routes were bypassed thus sealing their doom. At the Vulcan Mine near Iron Mountain while four men were coming up in the cage, a laborer on top lost control of a tram car and it tumbled down into the shaft, smashing and

---

[2]*Iron Ore*, June 25, 1887.

killing all four. At the Colby Mine in Bessemer, a fall of ground at the No. 4 shaft killed four men, and at the Cleveland Mine in Ishpeming, three miners, descending to work in the 530-foot-deep shaft were killed when the skip they were in overturned dumping them to their deaths in the inclined shaft.

Safety regulations and safety equipment in the early iron mines of the Marquette County area were limited in their effectiveness to prevent accidents and safeguard human lives. It was not until 1887 that a state law was put into effect requiring every county with an operational iron mine to hire a mine inspector who would check each mine at least once every 60 days. The county board of supervisors chose Anthoney Broad as mine inspector from a list of several who had applied for the new position. He was a well respected and experienced miner from Negaunee and in the post received a salary of $2,000 per year. At the time, some companies owned and operated several mines, with over 50 mines producing ore in Marquette County.

A week before Broad accepted the job, a local paper stated: "As long as mining is conducted accidents are bound to occur."[3] Later in December, shortly after two deaths had taken place in mines just examined, the same paper said: "The securing of such an office will count for little in reducing the chances of danger."[4] This was certainly no reflection on Broad's judgment, but merely an acceptance that nothing could improve the miner's plight.

Poor lighting in the underground workings was a main factor in the high accident rate which prevailed for many years in the area mines, but eventually the candles and lanterns gave way to the respected electric light. The Lake Superior Mine in Ishpeming was the first to test this new system of illumination on its properties.[5] Using the Brush system of generation, with power supplied to the dynamo from a coal-burning, steam operated engine, a few electric lamps were put to use in October, 1879. Even though the electric light system was basically more expensive than oil or candles, the test proved very satisfactory. Not many months later, in 1880, the Cleveland Iron Mining Company installed a Brush system which supported 18 lamps.

In December of 1882, the Republic Iron Company installed an elaborate lighting system that supported 16 lamps — each equal to 2,000 candles — in their huge Republic Mine at Iron City (Republic). A "straight line" 35 horsepower engine with cylinders of eight-inch bore and stroke of 14 inches, was installed. The engine was connected with shafts and belting to a Brush electric dynamo, which turned at 225 RPMs. Two single strand,

[3]*Iron Ore*, August 20, 1887.

[4]*Ibid.*, December 3, 1887.

[5]*Mining Journal*, October 4, 1879. (Later, on November 29, the paper noted: "From all accounts the Lake Superior mine is the first in the world which has been illuminated by electric light.")

quarter-inch copper wires, separated and insulated, were used to carry the current to the carbon lamps. The cost of the carbon was estimated at three-quarters of a cent per hour. When properly set, the carbon lamps would last overnight. The complete lighting plant, including the engine, cost $5,000.[6]

The amount of light given by these new lamps increased ore production to a certain extent and, with the savings on replacement of broken glass globes on oil lanterns, the electric light systems were financially justified.

SHIPPING

The *Detroit Advertiser* reported in May, 1848, that ". . . the steamer *Champion* arrived from Sault Ste. Marie on Saturday last, bringing several bars of Lake Superior Iron manufactured at the Jackson Iron Co.'s Works." From "several bars," the forge iron shipments increased and in May, 1852, 1,905 blooms of iron from the Marquette Iron Company were brought to the Sault on the steamer *Baltimore*, weighing an average of 126 pounds each and totalling more than 120 tons.

Later, when the Marquette County blast furnaces began their shipments, the tonnage of iron (pig iron) increased to thousands of tons each year. Though a very small percentage was used in the foundries in Marquette, the pig iron shipments for the first ten years of production in the Upper Peninsula were as follows:

1858	1,629 tons
1859	7,258 "
1860	5,895 "
1861	7,970 "
1862	8,590 "
1863	9,813 "
1864	13,872 "
1865	12,283 "
1866	18,437 "
1867	30,911 "

Shipments of pig iron from the Upper Peninsula blast furnaces increased to 84,489 tons, valued at $2,703,648, in 1874.

The first consignment of iron ore from the Marquette Range was from the Marquette Iron Company and put on board the steamer *Baltimore*, bound for the Sault, on July 7, 1852. After the long portage, the ". . . six bbls. of iron ore . . ."[7] were again loaded on a lakes vessel, destined for B. L. Webb of Detroit. The increase in individual carrier tonnage rose slowly and in October of 1868, the bark *Brightie* sailed from the Marquette harbor with 821 tons of ore on board — a record load.

[6]*Mining Journal*, December 2, 1882.
[7]*Ibid.*, December 14, 1872.

Steam barges towing schooners were a common sight on the lakes in the 1870s and 80s. Many barges such as the *Graves* with the schooner *Adams*, and the barge *Drake* with the schooner *Dot*, often lay in Marquette harbor waiting their turn under the pockets. The tug *Niagara* headed for the Sault in July, 1880, with eight ore-laden barges in tow, which was then the largest number ever towed on Lake Superior.[8] It was not unusual to see 35 or 40 vessels at anchor in the harbor at one time, and on September 3, 1882, 50 vessels waited for the trimmers to load the boats. The *Magnetic*, owned by the Republic Iron Company, steamed out of the harbor in August, 1880, with the largest load to date — 2,023 tons of ore. By the turn of the century, the big metal-hulled ore carriers were moving over 6,000 tons of ore each trip. The *Superior City,* owned by the Illinois Steel Co., cleared the Escanaba port in June of 1898 with 6,750 tons of ore, drawing 19 feet of water. The following month the *Presque Isle* loaded 6,031 tons of ore from the Lake Superior & Ishpeming Railroad dock in Marquette's Upper Harbor which set a new record on Lake Superior.

The process of loading the ships with wheelbarrows was eased to a great extent when the Cleveland Iron Mining Company built the first ore loading dock utilizing pockets in Marquette harbor. The pockets held from 50 to 80 tons of ore each and on June 22, 1859, the schooner *J. W. Sargent* received the first load of ore from the new facility.[9]

The railroads provided the link between the hundreds of mines and the ore docks. The Bay de Noc & Marquette Railroad in 1867 listed their rolling stock as ". . . eight locomotives, 500 freight cars, 4 passenger cars (pretty mean ones), 12 flat cars, and 4 horse cars . . . ."[10]

When the Marquette, Houghton & Ontonagon Railroad completed track laying to L'Anse on December 15, 1872, this opened another route for the iron ore to leave the Marquette Range, specifically from the Spurr Mountain and Michigamme mines. The railroad built an ore dock at L'Anse to facilitate handling of the ore. The dock at first had 40 pockets on one side, each holding 100 tons of ore, and plans called for 40 more pockets to be constructed on the other side before the next season began. On June 4, 1873, the schooner *Cambridge* was loaded with 23 cars of ore from the Michigamme mines, which was the first ore to be shipped from the L'Anse dock. By August, 1873, more than 30,000 tons of ore had been shipped from L'Anse.

The ore dock at St. Ignace, owned by the Detroit, Mackinac & Marquette Railroad, remained in service only a very short time with 60,159 tons of ore shipped through this facility in 1882.[11] This compared with

[8]*Mining Journal*, July 3, 1880.

[9]Hatcher, p. 74.

[10]*Lake Superior Mining & Manufacturing News,* May 15, 1867.

[11]Although it is documented in the Weekly Edition of the *Mining Journal* of February 23, 1884, and March 8, 1884, that the St. Ignace dock was being stripped of "materials" and "spouts," it was used later and in the October 11, 1890, issue called the "nearly idie dock at St. Ignace."

Escanaba's shipments of 1,622,654 tons for the year, and Marquette's 876,766 tons (L'Anse shipped 70,453 tons). High freight rates and the long haul from the mines did not justify the St. Ignace dock. The other lines (C&NW and MH&O) were making higher profits with cheaper freight rates.

The DM&M Railroad purchased all of the Cleveland Iron Mining Company's property in the city of Marquette in 1883, and abandoned plans for making St. Ignace an ore shipping port. Some cars of ore continued to go that way and they were moved across the Straits of Mackinac by the ferry *Algomah*, which made connections with the Michigan Central Railroad at Mackinaw City.[12] Work was started on the Marquette & Western Railroad, a subsidiary of the DM&M, on September 3, 1883. The line eventually ran from Marquette to the Winthrop Mine near Ishpeming, a distance of 17 miles. Two hundred men started work initially and soon 350 were on the job.

The DM&M began construction on its new ore dock in Marquette (for the Marquette & Western) in January, 1884, at about the same time it put a crew of men to work dismantling the St. Ignace dock. Fifty of the metal chutes from St. Ignace were shipped to Marquette in August. Edward Fraser, who had a sawmill on Cherry Creek, began delivering on a contract for 1,200,000 feet of lumber to be used in the new dock.[13] It was completed on July 2, 1884, and many car loads of ore from the Lake Superior Mine were run into her pockets (the dock had a capacity of about 9,500 tons).

The following day — Thursday, July 3, 1884 — was one long remembered in Marquette. Electric lights were turned on in the city for the first time! The electric-dyanmo used for supplying the power was owned by the DM&M Railroad and set up in its machine shop. From here, more than two miles of No. 8 copper wire, insulated with cotton asbestos, ran to the different properties. The DC circuit had one light at Ely's crossing, one at the railroad's new depot, two at Kaufman & Sons' store, one at the Front and Superior (Baraga Ave.) Streets crossing, four on the old Cleveland dock and three on the railroad's new dock, from which the wires returned to the dynamo. At Kaufman's store that night a large crowd gathered for this historic showing and the electric light was said to be "a big hit."[14] This day also saw the loading of the first boat at the new dock, the barge *Minnehaha* towed by the steam barge *Hiawatha*.

The only other port in the Upper Peninsula to build an ore dock before the turn of the century was Gladstone (other than the Huron Bay fiasco — an ore dock that was completed but never used) where the Minne-

---

[12]"It is getting slightly dangerous to cross the straits over the old trail. The heavy loads of ore which have been hauled across . . . have worn deep ruts, through which the water spurts up whenever the ice is jarred —" *Mining Journal* (Weekly Edition), March 7, 1885.

[13]*Iron Agitator*, April 19, 1884.

[14]*Mining Journal* (Weekly Edition), July 5, 1884.

apolis, Sault Ste. Marie & Atlantic Railroad began construction in January of 1889 on a 16,000-ton capacity dock. Dynamite was first used to break up the frozen ground for the necessary excavating which was quickly followed by pile drivers hammering down the timbers for the approach reaching out to the shore. The facility was soon completed and by the end of July of that year, it handled 25,561 tons of ore from the Marquette and Menominee Iron Ranges. For the year 1895, 190,310 tons of ore were shipped from Gladstone.

From six-ton ore cars to 75-ton "big blacks;" from a few hundred tons on the deck of a wooden sail boat, to a mammoth 32,000 tons steel-hulled carrier loading at Escanaba; from 500 tons of ore shipped in 1853, to 20 million tons of ore and high grade pellets shipped from the Upper Peninsula in 1974; from a railroad-owned dynamo serving two commercial customers, to a city based, municipal, private and public owned thermal generating plants, serving a greater part of the Upper Peninsula; the development of the region continues.

THE McCOMBER

IRON COMPANY.

DEALERS IN

Specular and Hematite Iron Ores,

FROM THEIR MINES AT

NEGAUNEE, MICHIGAN.

————•—————

SAMUEL L. MATHER, Pres't and Treas.
Cleveland, Ohio.
FRED. A. MOORE, Secretary,
Cleveland, Ohio.
JAY C. MORSE, General Agent,
Marquette, Mich.

From *Beard's Directory of Marquette County,* 1873.

GLOSSARY

1. **BANKED UP** — A term applied in blast furnace practice when the production of iron is stopped, but the furnace remains full of stock. In preparation for banking, a coke or charcoal blank is charged, consisting of from 60 to 200 tons of coke, and the blast is kept on the furnace until this blank is on the hearth, and in front of the tuyeres. The remaining stock in the furnace has reduced burden. When the coke or charcoal blank is in front of the tuyeres, the furnace is shut down, all slag drained out, and the furnace is then tightly sealed against air infiltration so that the coke or charcoal will not be burned away, but will be available for the start up of the furnace.

2. **BELL AND HOPPER** (Cup and Cone) — The charging apparatus at the top of a blast furnace, so devised that the raw materials can be fed into the furnace while the escape of gas is prevented.

3. **BLAST** — The current of air blown into the blast furnace through the tuyeres for the combustion of the fuel. The blast is usually preheated in which case it is known as hot blast or if unheated as cold blast.

4. **BLAST FURNACE GAS** — A gaseous fuel obtained as a by-product from the blast furnace. Although of low calorific value, owing to its high content of carbon dioxide, it is available in large quantities and is used in preheating the blast and for steam raising.

5. **BLOOM** — An intermediate product which has been rolled or forged down from an ingot and is destined for further working into bars, sheet, tubes, and forgings, etc. It is usually square in section and more than five inches square, smaller sizes being known as billets.

6. **BOSH** — The bosh is the widest part of the furnace and is also the area having the highest temperature.

7. **CHARCOAL KILN** — (a) High-quality wrought iron, charcoal having been used as a fuel in its production. (b) Pig iron which has been produced in a blast furnace using charcoal as fuel and, therefore, of unusually high purity.

8. **CHARGING** — The operation of introducing the raw materials into the furnace.

9. **CHILL CAST** — Pig iron cast into metal moulds or chills; if a machine is used the product is known as machine cast pig.

10. **CLOSED TOP FURNACE** — A blast furnace closed at the throat for the purpose of collecting gases.

11. **COKE** — The solid residue from the carbonization of coal after the volatile matter has been distilled off. It is used as a fuel.

12. **HEARTH** — In the blast furnace, that zone at the bottom of the furnace into which the molten pig iron trickles down and collects until it is tapped off.

13. **MERCHANT BAR** — The finished form of puddled bar after piling, reheating and rolling.

14. **MUCK BAR** — Bar rolled from a squeezed bloom.

15. **PIG** — A mass of metal cast in a simple shape, and subsequently remelted for purification, alloying, casting into final shape, or into ingots for rolling.

16. **PIG IRON** — Crude iron produced by the reduction of iron ore in a blast furnace and cast into pigs which are used for making steel, cast iron, or wrought iron.

17. **PIG BED** — Small excavations or regularly made open sand moulds in the floor, receive the molten pig iron from a blast furnace.

18. **PUDDLING** — A process, invented about 1890 and still in use, for the production of wrought iron. Pig iron is melted on the hearth of a small reverberatory furnace at a temperature above the melting point of pig iron but below that of wrought iron.

19. **SALAMANDER** — The mass of metal which is found below the hearth level of the blast furnace after the furnace has been blown out, and formed by the escape of molten iron through the hearth.

20. **SCAFFOLDS** — A term applied to obstructions formed in the blast furnace which impede the regular and even descent of the charge. Scaffolds may consist of displaced brickwork behind which deposition of carbon has occurred and agglomerated of slag, coke, iron oxides and reduced iron together with alkali silicate compounds, adhere to the refractory wall.

21. **SHINGLING** (Nobbling, Nobbing, Knobbling) — A stage in the puddling process in which the ball of iron is hammered or squeezed to form a bloom or billet of roughly rectangular slab, with the object of solidifying the metal and expelling, as far as possible, the intermixed liquid slag.

22. **SLIPS** — Irregularities in blast furnace practice caused by wedging of stock in the upper part of the stack.

23. **TUYERE** — A nozzle through which air is blown and distributed through the blast furnace; the tuyeres are situated below the bosh and vary in number according to the size of the furnace.

24. **WROUGHT IRON** — The chief characteristic of wrought iron is that the temperatures employed in its production are too low to render it fluid and its condition is never more than pasty or semi-fused. Hence it contains an appreciable quantity of slag. On hammering, the metal granules are elongated and more or less welded together whilst much of the slag is squeezed out, but some remains intermingled with the iron in thread-like form, thus giving the characteristic fibrous structure of wrought iron.*

*Terms from *An Encyclopaedia of the Iron & Steel Industry*, by A. K. Osborne, (Philosophical Library, Inc., 1956.)

BIBLIOGRAPHY

The A B C of Iron and Steel. Edited by A. O. Backert, Cleveland, Ohio: The Penton Publishing Company, 1925.

Benison, Saul. *Railroads, Land and Iron: A Phase in the Career of Lewis Henry Morgan.* Ann Arbor, Mich.: University Microfilm, 1954.

Boyer, Kenyon, *Historical Highlights.* "The Carp River Blast Furnace," radio script of the Marquette County Historical Society, Marquette, Mich. Volume XIV, No. 247.

Brinks, Herbert John. *Peter White: A Career of Business and Politics in an Industrial Frontier Community.* Ann Arbor, Mich.: University Microfilm, 1965.

Fisher, Douglas Alan. *The Epic of Steel.* New York: Harper & Row, 1963.

Geological Survey of Michigan. New York: Published by authority of the Legislature of Michigan, Julius Bien, 1873.

Hatcher, Harlan. *A Century of Iron and Men.* New York, Indianapolis: The Bobbs-Merrill Company, Inc., 1950.

Hollbrook, Stewart Hall. *Iron Brew.* New York: The McMillan Company, 1939.

Lake Superior Mining Institute. Volumes II, IX, XIX, XXI, XXV. Published by the Institute. Printed by Darius D. Thorp, Lansing, Mich., and by the presses of the *Iron Ore*, Ishpeming, Mich.

Michigan Pioneer Collections. Lansing: The Pioneer Society of Michigan. 1885, Vol. VII.

Newitt, George A. *Mines and Mineral Statistics - State of Michigan.* Lansing: Robert Smith & Co., 1896.

Osborne, A. K. *An Encyclopaedia of the Iron & Steel Industry.* Philosophical Library, Inc., 1956.

Schallenberg, Richard H. *Innovation in the American Charcoal Industry, 1830-1930.* 1970.

Swank, James M. *History of the Manufacture of Iron in all Ages.* Philadelphia: The American Iron and Steel Association, 1892.

Williams, Ralph D. *The Honorable Peter White.* Cleveland, Ohio: The Penton Publishing Company, 1907.

The Daily Mining Journal. Marquette, Mich., 1893, 1904, 1910.

The Iron Agitator. Ishpeming, Mich., 1879-1886.

The Iron Ore. Ishpeming, Mich., 1887-1900.

Lake Superior Journal. Marquette, Mich., 1855-1863.

The Lake Superior Mining & Manufacturing News. Negaunee, Mich., 1867, 1868.

Lake Superior News and Mining Journal. Sault Ste. Marie and Detroit, Mich., 1847-1855.

Lake Superior News and Miners' Journal. Copper Harbor, Mich., 1846.

The Mining Journal (weekly edition). 1869-1910.

Negaunee (Mich.) *Iron Herald*, 1884-1885.

Negaunee Review. 1870.

UPPER PENINSULA FORGES AND BLAST FURNACES BEFORE 1900

FORGES	START UP	LAST KNOWN OPERATION
Carp River Forge	Feb. 10, 1848	June, 1857, last reference
Marquette Iron Co.	About July 6, 1850	Burned, Dec. 14, 1853
Forest Iron Co.	July, 1855	May, 1862, last reference
Collins Forge	August, 1855	Converted into experimental blast furnace, Jan., 1858

BLAST FURNACES	START UP	LAST KNOWN OPERATION
Collins Experimental Furnace	Jan. 21, 1858	Abandoned Jan. 25, 1858
Pioneer No. 1 Stack	April 26, 1858	Major breakdown in June, 1892, never started again
Collins Furnace	Dec. 13, 1858	Exhausted fuel supply in May, 1873
Pioneer No. 2 Stack	May, 1859	Blown out May 18, 1893, completely worn out
Northern Furnace	Summer, 1860	Made iron in 1910 for a short while
Bancroft Furnace	May, 1861	River overflowed May, 1876, destroyed flumes, shutdown
Morgan Furnace	Nov. 27, 1863	Blown out December, 1876
Greenwood Furnace	June, 1865	Bad roads, exhausted ore and fuel, shut down April 9, 1875
Michigan Furnace	Feb. 10, 1867	Blown out for repairs Jan 12, 1875, never restarted
Champion Furnace	Dec. 4, 1867	Burned, April 9, 1874
Jackson No. 1 Stack (Fayette)	Dec. 25, 1867	Company closed down, Dec., 1890
Schoolcraft Furnace (Munising)	June 28, 1868	Blew out, Nov., 1877
Deer Lake No. 1 Stack	About Sept. 1, 1868	Shut down, late fall of 1891
Bay Furnace No. 1 Stack	March 5, 1870	Burned, May 31, 1877
Jackson No. 2 Stack (Fayette)	May 2, 1870	Company closed down, Dec., 1890
Peat Furnace (Excelsior)	March, 1872	Shut down Nov. 5, 1897, hurting Ishpeming's water supply
Marquette & Pacific	July 13, 1871	Fall of 1881 or early in 1882
Grace Furnace	Dec. 10, 1872	Out of blast March 24, 1874, never used again
Bay Furnace No. 2 Stack	Dec. 15, 1872	Burned, May 31, 1877
Escanaba Furnace (Cascade)	Early in 1873	Blown out late in 1874, dismantled in 1879
Menominee Furnace	July 8, 1873	Possibly in late 1883
Deer Lake No. 2 Stack	Jan., 1874	Shut down, late fall of 1891
Cliffs Furnace	March 14, 1874	Abandoned about 1877
Carp River Furnace	April 26, 1874	Shut down, 1907
Martel Furnace	Aug. 14, 1881	Blown out 1903, fire destroyed stock house
Vulcan Furnace	May 21, 1883	Renovated 1902, possibly 1945
Iron River Furnace (Gogebic)	Feb. 2, 1886	Exhausted ore and fuel, May, 1888
Weston Furnace	April 30, 1891	Closed down, 1922
Gladstone Furnace	April 16, 1896	Closed down, 1922

UPPER PENINSULA FURNACE PRODUCTION FIGURES

(In Tons)

Bancroft[2]	55,608
Bay Furnace[1] (two stacks)	50,706
Carp River[2]	83,500
Champion[1]	31,048
Cliffs[1]	8,209
Collins[1]	41,997
Deer Lake[2] (two stacks)	93,579
Escanaba[1]	8,650
Excelsior[2]	68,634
Fayette[2] (two stacks)	229,288
Gogebic[2]	3,700
Grace[1]	11,346
Greenwood[1]	40,202
Manistique[2]	150,904
Martel[2]	58,349
Marquette & Pacific[1]	41,857
Menominee[2]	59,553
Michigan[1]	41,531
Morgan[1]	57,573
Munising[1] (Schoolcraft)	28,312
Northern[1] (15,059 to 1881, estimate for 1891 and 1910 operation: 15,000)	30,059
Pioneer[2] (two stacks at Negaunee and the Pioneer at Gladstone through 1902)	637,299
Vulcan[2]	73,829
	1,905,733

[1] *Mining Journal* (Weekly Edition), January 15, 1881.
[2] *Transactions of the Lake Superior Mining Institute*, Vol. IX, 1903. p. 71.

INDEX